The Moral Meaning of Nature

The Moral Meaning of Nature

Nietzsche's Darwinian Religion and Its Critics

PETER J. WOODFORD

The University of Chicago Press
Chicago and London

The University of Chicago Press, Chicago 60637
The University of Chicago Press, Ltd., London
© 2018 by The University of Chicago
All rights reserved. No part of this book may be used or reproduced in any manner whatsoever without written permission, except in the case of brief quotations in critical articles and reviews. For more information, contact the University of Chicago Press, 1427 E. 60th St., Chicago, IL 60637.
Published 2018
Printed and bound by CPI Group (UK) Ltd, Croydon, CR0 4YY

27 26 25 24 23 22 21 20 19 18   1 2 3 4 5

ISBN-13: 978-0-226-53975-1 (cloth)
ISBN-13: 978-0-226-53989-8 (paper)
ISBN-13: 978-0-226-53992-8 (e-book)
DOI: https://doi.org/10.7208/chicago/9780226539928.001.0001

Library of Congress Cataloging-in-Publication Data

Names: Woodford, Peter J., author.
Title: The moral meaning of nature : Nietzsche's Darwinian religion and its critics / Peter J. Woodford.
Description: Chicago ; London : The University of Chicago Press, 2018. | Includes bibliographical references and index.
Identifiers: LCCN 2017039546 | ISBN 9780226539751 (cloth : alk. paper) | ISBN 9780226539898 (pbk. : alk. paper) | ISBN 9780226539928 (e-book)
Subjects: LCSH: Nietzsche, Friedrich Wilhelm, 1844–1900. | Overbeck, Franz, 1837–1905. | Simmel, Georg, 1858–1918. | Rickert, Heinrich, 1863–1936. | Life. | Evolution (Biology)—Moral and ethical aspects. | Evolution (Biology)— Religious aspects. | Philosophy and science—Germany—History. | Philosophy, German—19th century. | Philosophy, German—20th century.
Classification: LCC B3318.N3 W66 2018 | DDC 193—dc23
LC record available at https://lccn.loc.gov/2017039546

♾ This paper meets the requirements of ANSI/NISO Z39.48-1992 (Permanence of Paper).

Contents

Preface vii

Introduction 1

1 Friedrich Nietzsche: A Darwinian Religion 25

2 Franz Overbeck: The Life History of Asceticism 52

3 Georg Simmel: Evolution and the Self-Transcendence of Life 79

4 Heinrich Rickert: The Autonomy of Agency and the Science of Life 105

Conclusion 132

Acknowledgments 151
Notes 153
Bibliography 171
Index 179

Preface

Any invitation to work through the nuances of past thinkers—many of whom are unknown to contemporary readers—requires something of a defense and, perhaps even, an exhortation. So, here it goes. It is perhaps not an overstatement to say that most philosophers today see the central task of philosophy to be one of establishing what the advancing sciences, especially neuroscience, cognitive science, and evolutionary biology, tell us about the topics that have animated Western philosophical and theological traditions—the topics of mind, agency, knowledge, ethics, and religion. Those provoked to reflect on these topics today cannot ignore the sciences, yet it is also fair to say that it is far from clear just what science tells us, or can tell us, without veering beyond bare data into philosophical assumptions that we bring to our investigation. For historians, this is of course not a new state of affairs. It is only since the seventeenth and eighteenth centuries that distinctions between separate domains of science, philosophy, and religious thought began to emerge. Indeed, it is a remarkable fact that the term "scientist," used to designate a definable professional and social role with a distinct aim and set of interests, was coined only in 1833 by the English philosopher and theologian William Whewell. Since the origins of modernity in the West—whenever that was—and the new understandings of nature that underpinned the scientific revolutions of the seventeenth and eighteenth centuries, students of nature, philosophers, and religious thinkers have been interlocked in a sort of scrum. The territory, legitimacy, and even the identities of each of these complex spheres was never a given, brute fact but has been constructed through debate over fundamental questions, new knowledge-gathering techniques, new conceptions of natural processes, and the challenge of relating the "new" to the "old." The motivation to understand the negotiation between these spheres, especially in response to new understandings of nature, animated my writing of this book.

What Darwin was and has become for discussions of religion and science in the Anglo-American contexts, Friedrich Nietzsche has been on the Continent. Nietzsche represented starkly for a generation of German philosophers the promise and peril of Darwinian, or even more broadly naturalistic, views for ethics, metaphysics, and the philosophy of religion. Indeed, Nietzsche is still a live resource for a stunning variety of contemporary disciplines, and his work has increasingly drawn the attention of philosophers working in both the Analytic and Continental traditions. This can—I contend—be explained in large part by the fact that Nietzsche's positions in various areas were undergirded by a general aim to translate human philosophical and religious concerns into the framework of an evolutionary picture of life. The other three thinkers in this study are examples of those who confronted the philosophical implications of Darwinian and biological thinking through their reading of Nietzsche.

Nietzsche appeared to have presented even more trenchantly the animating impulses behind traditional religious beliefs and aspirations than the allegedly more "superficial" "English psychologists" like Darwin (whom he criticized), and so his brand of naturalism appeared to present a much more compelling account of what evolution meant for philosophy. But, this book will stress, Nietzsche was driven by existential, and I will even go so far as to say religious, aspirations that were foreign to Darwin. This book reconstructs how Nietzsche negotiated the relationship between religion and science, how three of his most immediate critics steeped in the same tradition of German thought—Franz Overbeck, Georg Simmel, and Heinrich Rickert—responded to the same concerns, and how evolutionary conceptions of nature and biological life informed this debate in each case. These thinkers lived in a time during which philosophy was dominated by the question of its task and in relation to the advancing sciences, and this problem shaped discussion of Darwin. Understanding the negotiation of science, philosophy, and religion around the topic of evolution in this philosophical moment can, I hope, inform our own understanding of the complex entanglement between these areas today.

With this exhortation, it will be clear that this book spans two distinct sorts of interest that might be thought to be in tension, or at least not obviously related. This book was written to satisfy both historical and philosophical interests. Yet it is often thought that while historians are less concerned with the relevance of the past to the present, philosophers (and perhaps scientists as well) are less concerned with the past merely for its own sake, unrelated to our own task of rigorously articulating and defending understandings that we take to be the best. The following study examines the intersection of science, philosophy, and religious thought through the case of a unique debate that arose in Germany in the wake of Darwin's impact, and in carrying it out I set myself the difficult challenge of satisfying both interests. It is my confession at the outset that I am a hybrid, and that the mixture of history and philosophy that I aimed to achieve is of a particular sort.

In order to set the right expectations of what this book is and what it is not, let me describe in greater detail the particular blend of history and philosophy that I aimed to achieve. There is no grand thesis about the philosophy of history or the history of philosophy behind the method and aims of this study of four past thinkers. Nonetheless, thinkers of the past can speak to the present, and they can do so in a number of ways. They can offer us specific positions and ways of seeing the world that we might adopt wholesale and attempt to defend against objections; they can point us to genuine problems that we have overseen, and that need to be understood; they can manifest unresolved tensions between conflicting views that we can then aim to resolve in a more satisfactory way; and, they can demonstrate a guiding concern or motivation, an interest in the subject matter, that inspires us and helps us understand the motivations behind their, and our, intellectual efforts. Of course, these ways of speaking to the present—not intended to be exhaustive—presuppose that there is some common affinity between "us" and "them," that we can see ourselves as sharing problems and aims. The figures in this study were chosen on the basis of such an affinity and shared set of problems, which I for the moment I accept as a brute "given" of modern Western thought, at the very least. Of course, one might explain this given affinity through historical and cultural genealogy. Or it might instead be explained through features of a common and general human condition that we share—perhaps due to the inevitable questions a self-conscious life-form and product of evolution runs into as it is thrown into reflection upon itself, the world around it, and the origins of both. While I tend toward the latter, for the purposes of explaining the motivations behind this book it suffices for now to take this affinity as a matter of fact that might be explained either way.

The reason these thinkers have been chosen is that they were united in their approach to the relation between philosophy, religion, and science by a common interest in the question of what meaning an evolutionary view of nature had for human practical life—what satisfactions humans seek, what ideals they strive to realize, and how they understand their own agency. While discussions of evolution and religion that took place in the English context in wake of Darwin focused mostly on the compatibility between Darwin's thesis and specific religious beliefs regarding creation, human origins, providence, and arguments for the existence of God, the discussion that Nietzsche inaugurated focused on the sources of values and the normative validity of these values within an evolutionary conception of life. It is my contention, and the thesis of the book, that because of their focus on the problem of value, each of these thinkers displayed an interest in science not primarily for the purely theoretical end of describing and explaining the causes of natural events, but also for an existential interest in what I am calling the *meaning* of nature. Their mediation between religion and science was shaped by the attempt to understand the implications of the scientific picture of nature for decidedly non- and extra-scientific questions.

While the criterion of selection used to select just these figures has been driven by philosophical interest in this theme, the aim of the book is not primarily to advocate for or defend any particular position. The method of the book is historical analysis and rational reconstruction. My goal in writing it was to provide enough historical context on each thinker to show that the strategies they adopted with respect to the negotiation between philosophy, science, and religion were embedded within and arose out of a distinct intellectual lineage that developed a unique vocabulary and set of concerns. Of course, the inevitable danger that such a bridging of historical and philosophical interests meets is that historians may find too little attention to historical context and broader social and cultural implications, while philosophers may not find enough analysis of arguments to determine whether or not these positions are defensible against potential objections. In each chapter, I reconstruct each thinker's position on the relationship between science and religion and show how an interest in the meaning of nature for human ideals and values drove their reflection on this relationship. In this sense, the book is more historical than philosophical. But, in that the thinkers and subject matter have been chosen particularly because they were focused on problems that have contemporary resonance, the interest that it satisfies is not purely historical. In the conclusion, I discuss what I think are the most important lessons of these debates for contemporary philosophy. I leave it open and would consider it an accomplishment if any readers feel a pull to develop Nietzsche's position, or that of any of his critics, and to defend them against the objections they raised against one another or that they might meet today. Sensitivity to the importance of historical context alone should not, of course, rule out the possibility that past thinkers got things right.

Introduction

> When we speak of values, we speak under the inspiration, under the optic of life: life itself forces us to establish values; when we establish values, life itself values through us.
> FRIEDRICH NIETZSCHE, *Twilight of the Idols*

> The idea of a science is always the concept of a task to be carried out.
> HEINRICH RICKERT, *The System of Philosophy*

Evolutionary biology has come a long way since Charles Darwin, but despite widespread agreement on the basic picture of life evolving from simple to complex through a process of descent, modification, and selection, the questions that it generated for ethicists and philosophers of religion have changed remarkably little. Our understanding of the evolution of living things over the scale of the earth's deep history has shown the origins of the human species, the dynamic processes that shape the enormous diversity of forms of life, and the range of selective pressures that challenge them to survive and reproduce successfully. Though still controversial, there is growing literature that acknowledges the power of evolutionary principles for understanding the origins of religions, of ethical values, and of many further aspects of human culture and social life as well. Yet despite incredible advances, the wider implications of an evolutionary understanding of life's history and of the origins of humanity are as hotly debated now as they were in Darwin's time, and, for many, they have not yet been fully settled.

Of course, Darwin's theory of evolution as a process of descent with modification, driven by the dynamics of natural selection, immediately sparked intense debates among scientists, philosophers, and theologians at its own origin. In the English context, the most sophisticated of these debates centered on the implications of Darwin's theory for the then formidable tradition of natural theology, most forcefully articulated in William Paley's classic work from 1802 by the same name. While the tradition of natural theology has now come to be recognized as complex, one of its principle aims was to defend the reasonableness of the inference from observable features of the natural world to the existence of a God, conceived as the supreme architect of nature and source of apparent order, design, and purposiveness. With respect to the living world, Paley argued that the most salient of its features was the *teleological*, or purposive, structure and behavior of living things—the way that living things might be seen as analogous to a watch with a unified purpose through which it, and the interaction of its constituent parts, could be understood.

Darwin's theory made possible a revolutionary approach to the problem of the teleological design of the living world, most apparent in the ways in which living things were so extraordinarily adapted to fit the environments in which they had to survive and make their living. Darwin's concept of natural selection explained this design as the product of a slow, gradual, cumulative, and stepwise process of variation, inheritance, and modification that took place over billions of years.[1] One of the most incredible features of Darwin's theory for philosophy that came to be seized upon by later generations—up to the present—was its apparent ability to account for the purposive structure of living things without invoking an intentional designer or cosmic mind outside of blind and random processes that produce biological variation and alterations of structure and function.[2] This was something almost unimaginable just seventy years before Darwin, as Immanuel Kant famously evinced through his claim that there can be no "Newton of the blade of grass" who is capable of explaining organic life according to blind, mechanical processes, and it appeared to make a purely mechanistic, materialist understanding of life possible in a way it could not have been before.[3] Yet, while the implications of Darwin's theory of evolution by natural selection for the traditional of natural theology were central to and have come to define debates over science and religion that focus on evolution today, the problem of whether or not life's emergence and evolution is blind and random does not capture the full extent of the challenges that evolutionary theory raised for philosophy and religious thought then or now.[4]

In Germany, some of the earliest and most influential appropriators of Darwin seized upon a different set of issues that were understood to be just as culturally, and even existentially, urgent. Instead of the pressing puzzle of how a machinelike design can be produced without an intellect as its cause, these thinkers saw life and its evolution as evidence of a creative power inherent in nature. The pantheistic and Spinozist strain of German romantic and Idealist philosophy of nature in thinkers like Herder, Goethe, Schelling, Schleiermacher, and Humboldt, which deified nature and its creativity, had already warmed late nineteenth-century German thinkers to an immanent conception of the divine, manifested in the creativity of nature, against the notion of a transcendent, intentional creator *ex nihilo*.[5] Instead of puzzling over the designlike character of living things, these thinkers tried to sort out the philosophical and theological implications of Darwin's work by asking what meaning this new understanding of biological life in nature might have for human life, understood as life that was oriented by the attempt to realize values. What did this view of nature mean for the peculiarly human existential "lot," the human condition of having to choose, live out, and pass on a set of ideals that declared not how nature *is*, but how it might be made to be? These thinkers debated whether or not values, ends, and ideals fit into life itself and were encoded in the processes that generated it. They aimed to assess how the values and projects that were at the basis of European culture were matched against the driving power behind nature's own creative process.

This book concentrates on a specific set of thinkers in these early debates in Germany that followed upon the heels of Darwin's theory of evolution. It is about what motivated them, how they reconciled this new picture of nature with their philosophical and theological heritage, and how their debates over biology and human behavior generated a unique approach to the relationship between religions and the sciences. This study returns to a philosophical context in which reflection on the right relation between philosophy and biology was focused on a central and paradoxical hermeneutic question: What is the moral *meaning* of nature? Of course, it is important to point out that perhaps only human beings, so far as we know, are the ones for whom the meaning of nature is a question. And, indeed, understanding what this question itself means is something that must come out in the course of this study. The thinkers in this book were all driven by the attempt to draw out the implications of marrying two worlds, the world of nonhuman living things, on the one hand, and the world of human religious and ethical striving, on the other. The question of the moral meaning of nature addressed the significance of nature's own "ends"—if there were any—for reflecting on the ends and ideals that were presupposed in, and that guided, human life. For the group of thinkers I examine in the following chapters, the problem of value, rather than that of design, became central to sorting out the philosophical and theological implications of Darwin and evolutionary biology, and this historical investigation is intended to inform our continuing meditation on what evolution means for human ethical and religious values.[6]

Who are these thinkers, and why have they been selected? The thinkers in this study have been selected because they belong to a distinct lineage in which the question about the meaning of Darwinian nature *for life*—or more pointedly, for *our life*—became central for mediating the relationship between science and religion. The thinker who explicitly raised this question in a way that shaped German debates over evolution, ethics, and religion was less Darwin than it was Friedrich Nietzsche. Nietzsche's appropriation of biological thought established new terms and a new terrain for debates over the relationship between religion and science. It did this largely by rejecting what philosophers at the time considered to be the overly reductive and mechanistic (in senses to be defined more precisely in the following chapters) conceptions of life and by depicting human agency as an instance of a more general form of agency possessed by all biological life, out of which the specifically human form emerged. This book reconstructs Nietzsche's formulation of and answer to the question of the moral meaning of nature through his unique appropriation and critique of Darwin. However, the goal of this book is not only to present Nietzsche's negotiation of science, philosophy, and religion around his concepts of life and value. It also explores the problem of the philosophical significance of evolutionary biology through an examination of early, critical reactions to Nietzsche, Darwin, and Nietzsche's Darwin.

Two of Nietzsche's earliest acolytes adopted his terms and problems, but they also became trenchant critics. They each took Nietzsche's Darwinian philosophy of life in different directions that were related to their own disciplinary concerns. The first, Franz Overbeck (1837–1905), was Nietzsche's close friend and former neighbor in a house that they both lived in as they started their academic careers at the University of Basel in the beginning of the 1870s. Overbeck famously took care of Nietzsche's finances during his nomadic years of prolific writing; he rescued him from Turin after his breakdown in 1889 and brought him to the care of his sister and mother; and he battled the notorious attempts by Nietzsche's sister to manipulate her brother's legacy for German nationalist aims in the years following his tragic mental decline. Yet Overbeck was a formidable scholar; he was a historian of the New Testament and Christian origins and was housed in the Theological faculty in Basel for his entire career, until his death in 1906. Overbeck wrote penetrating scholarly essays on early Christianity that were informed by the philosophical lessons he took from Nietzsche's philosophy of life and the discussions of evolution taking place in theology. However, Overbeck was also critical of Nietzsche on important and central points, and as a historian, he was able to assess the historical accuracy of Nietzsche's compelling, but ultimately speculative, Darwinian just-so stories about the origins of morality and Christianity. Overbeck applied the concept of life to the actual history of Christianity and used it to address a foundational problem concerning the status of the academic discipline of theology in relation to the sciences. He too came to understand that the question of the moral meaning of nature is as decisive for assessing the relationship between science and religion, which he understood in terms of a wider conflict between the pursuit of scientific knowledge and other needs of life.

Georg Simmel (1858–1918) was also an early appropriator and critic of Nietzsche, especially of Nietzsche's concept of life and his appropriation of Darwinian ideas. Simmel began his academic career under the influence of a naturalistic school of social psychology in Germany called *Völkerpsychologie* (the psychology of peoples). This school aimed to explain social structures, shared communal norms, and the values that tied groups together according to underlying natural drives that individuals within a group shared in the context of common ecological settings. Simmel was the first to hold lectures on Nietzsche's work in a German academy in 1902 and was one of the earliest of Nietzsche's interpreters to focus on Darwin as a key influence. Simmel was an insightful and sympathetic reader of Nietzsche, but he was also critical of what he took to be the decisive principle of Nietzsche's thought—namely, the concept of life. Simmel was skeptical of Nietzsche's attempt to place the life sciences in the service of reflection on normative ethical and religious interests, yet he, too, later adopted and tried to improve upon Nietzsche's notion of life in response to neo-Kantian critics who found its theory of value and agency inadequate. In so doing, he transformed Nietzsche's concept of life through his own account of what it meant to view life as a creative power behind the worlds of both nature and culture.

Finally, I turn to the most scathing critic of all the Darwinian "trends" that had allegedly "infected" fin-de-siècle philosophy: the neo-Kantian philosopher Heinrich Rickert (1863–1936).[7] Rickert took Nietzsche and his heirs to represent the most consummate and powerful form of philosophical Darwinism, but one that was nonetheless still mired in all of its errors. He was the culminating representative of the Southwest, or Baden, School of neo-Kantian thought that was centered around Heidelberg and Freiburg. Rickert labeled all naturalistic, vitalist, and Darwinian philosophers of his day *Lebensphilosophen* (Life-philosophers), and in 1920 he published a book entitled *The Philosophy of Life: A Presentation and Critique of the Philosophical Fashions of Our Time* that attempted to put the nail in the coffin of all attempts to draw philosophical lessons from the life sciences. While Darwin, Nietzsche, and biological theory seemed to offer the possibility of understanding values as products of basic needs and drives present in living things generally, Rickert argued that the concrete content of cultural values embedded in the major spheres of culture—namely, religions, the sciences themselves, communal life, politics, and aesthetic enjoyment—definitively left life behind. The values that constituted these spheres regulated, extended, and often conflicted with tendencies that could be considered *merely* biological.

Why Consider Nietzsche and His Critics Today?

This return to a bygone set of debates over philosophy, biology, and religion is motivated in part by the fact that the questions these thinkers were asking are still alive for us today. The way in which the thinkers in this study understood the difficulties involved in bringing together the problems of philosophy, the study of religion, and biology, while not necessarily resolving them for us, can help to inform and broaden our own discussions of these issues. They can also help us see how the debate about the relationship between science and religion that emerged out of reflection on evolution and the life sciences was very different from the ones that often occur today. Before returning to Nietzsche and his critics, I would like to draw out some of the unique aspects of their discussion in relation to contemporary research concerning evolution, values, and religion.

In the contemporary science of evolution, questions about evolution and value are often posed in terms of *moral* value and, more specifically, in terms of how natural selection could have favored so-called prosocial or even altruistic attitudes and behaviors. This question has become so interesting, and so pressing, because of assumptions about the nature and extent of competition in the struggle for life that were originally made by Darwin himself. Relying on the economist Thomas Malthus, Darwin reasoned that population increases would lead to a scarcity of resources needed to survive and reproduce, and this would lead to competition. Only those most fit to obtain vital resources—that is, food and mating opportunities—would manage to pass on their traits to the next generation. The so-called puzzle of the evolution of human morality lay in the question of how traits that lead humans, or any other animal, to provide aid to others, often at costs to themselves, could evolve, given that competition over resources was fundamental to how natural selection would shape behavior. Natural selection, it seemed, could only favor those who outcompeted others to obtain the resources necessary to live and to pass life on to offspring, so how could behaviors that lead organisms to cooperate, instead of compete, ever evolve?

In the last decade of the nineteenth century, Peter Kropotkin objected to what he saw as undue emphasis on competition that resulted in the neglect of cooperation in the study of animal behavior, but it took some time until this problem appeared to be definitively solved.[8] Today, the solution is credited to William Hamilton's concept of "kin-selection" and, further, to the advent of evolutionary game theory. Hamilton showed how cooperative and altruistic behaviors can be favored by natural selection if these behaviors have a high probability of benefiting relatives, because relatives were likely to share the same heritable basis of these cooperative or altruistic traits. The idea was that genetic or other heritable components that are involved in cooperative behaviors could spread either directly, through an individual's reproductive success, or indirectly, through the reproductive success of others who share those heritable components. Evolutionary game theorists today also stress that in the context of social behavior, in which multiple individuals interact in ways that affect each other's ability to survive and reproduce, helping can emerge as a highly successful pattern of behavior under many conditions.[9] These theoretical advances show that it simply is not true that individualistic opportunism is the only, or the best, route to the so-called ultimate evolutionary aim of survival and reproductive success.[10]

Recent work on the evolution of human morality and religion has focused on the role these play in sustaining cooperative behavior and on the adaptive benefits of the human tendency to cooperate with others for the sake of obtaining material and reproductive resources.[11] The evolutionary utility of interest in the welfare of others is understood through the benefits to the reproductive success of genes, or genetic ensembles, that shared action and mutual aid can provide. Since it is argued that the disposition to care about the welfare of others underlies the development of moral norms, and religious beliefs function to reinforce moral norms, the evolution of morality and religion is explained by whatever benefits to reproductive success cooperation tended to confer upon individuals who adopted moral or religious attitudes or lived in societies that instituted such norms. So, contemporary evolutionary theory understands these benefits of cooperating in a very narrow sense: these are the benefits of reproductive success and survival, but survival only insofar as it leads to reproductive success. These are the ultimate measures of cost and benefit, the criteria of "good" and "evil"—to put it figuratively—from the perspective of evolutionary theory.

While this work is clearly important and has led to increasing insight into the evolutionary process and its products, the narrow focus on the mechanistic problem of how heritable traits spread fails to bring into view important issues that were central to the debates between Nietzsche and his critics, and that ought to attend discussion of evolution, value, and religion today. The thinkers in the following study add a unique perspective to contemporary research because they viewed values embedded in ethical and religious systems, and indeed also in politics, art, and even science, as introducing aims for action that could not be made sense of narrowly in terms of biological fitness. Biological fitness today refers to measures of gene copies that make it into future generations. But these arenas of cultural value introduced values with content that was indifferent to this process, values that made claims on human action *for their own sake* and not because of their evolutionary effects. They thus introduced different criteria for what counts as success in human action, and these criteria were intrinsic and constitutive of the various arenas of human activity that they guided.

The concept of life became so important in the tradition recounted here because values seemed to be entwined with the activity of living things. Values only seemed relevant to living processes, because it was through these processes that they emerged, and it was these processes on which they appeared to make a claim. The aim of bringing biological theory into conversation with philosophy was to determine to what extent values could be understood in relation to the characteristic features of living processes. In bringing philosophy into conversation with biological theory, the thinkers in this book blurred the strict division between science and value; they wanted to know whether or not science could be put into the service of life. For this reason, their inquiry was not only about the causal history of values and the mechanisms of transmission; it was also a meta-ethical, metaphysical, and even existential investigation into the nature of value, into the sources of value in nature, and into how one might best understand the activity of valuing in the context of an evolving, Darwinian natural world. More will be said about these questions in the rest of this introduction.

Contemporary work in the philosophy of science and highly nuanced discussions of value and normativity in Analytic meta-ethics have addressed these issues with an astonishing degree of rigor and detail in the last decades.[12] In the tradition of Analytic meta-ethics, concerns over the relationship between descriptive and normative claims are often traced back to G. E. Moore's famous critique of ethical naturalism, which he worked out in *Principia Ethica* in 1903. Moore's famous "open-question" argument stated that goodness cannot be conceived as a natural property because any set of descriptions about the natural properties of various entities always left the question open as to whether these properties individually or collectively were indeed good. Moore's argument reinforced the idea that there is a "naturalistic fallacy" involved in the logical derivation of evaluative or normative "ought" claims from the description of natural occurrences, states of affairs, or properties of physical objects. Yet Moore's worries themselves retrieved a foundational issue of moral philosophy in the tradition of British Empiricism, especially from David Hume. Moore's worry about the naturalistic fallacy and its source in Hume is important background to the discussion to be recounted here, since Immanuel Kant's moral philosophy was a direct response to Hume, and Kant set the agenda for Nietzsche and his critics, who were negotiating between a growing empiricism and positivistic stance in the sciences and the heritage of rationalist, transcendental philosophy. Moore's critique of ethical naturalism thus occurred at the same time that thinkers from what has now become the so-called Continental tradition in Germany, such as the Marburg neo-Kantians Hermann Cohen and Paul Natorp, Edmund Husserl, and the Baden neo-Kantians Wilhelm Windelband and Heinrich Rickert, were developing their own criticisms of naturalist approaches to normativity. The problem of naturalism and value is at the heart of Nietzsche's critical appropriation of Darwin at this moment—a moment in philosophy during which the role of philosophy next to the sciences was the principle issue of the day. The so-called Continental thinkers in this study understood the problem of value and the transcendental question of the sources of normativity that was central to Kant in a way that took it beyond the sphere of ethics or questions of the moral good into all areas of cultural and *geistig* activity, including both the philosophy of science and the philosophy of religion.

In popular culture today, discussions of the conflict between science and religion have largely been shaped by the polarized writings of the so-called New Atheists—Richard Dawkins, Sam Harris, Christopher Hitchens, and Lawrence Krauss—and the reactions to them. While these thinkers have reinvigorated debates over science and religion in the popular sphere, philosophical assumptions about just what science and religion *are* that underlie these discussions often go unacknowledged and, where they are acknowledged, they are not very sophisticated. Science and religion are contrasted crudely in terms of "faith" versus "fact."[13] The questions of whether meta-ethical and metaphysical lessons can be drawn from the sciences and from nature itself, and which lessons these should be, do not come into view. And yet, if the lessons of the thinkers in this study are to be heeded, these issues continue to be crucial for understanding how the two "magisteria" of science and religion relate to one another.[14] This book shows how a collection of thinkers conceived the relation between science and religion by focusing on the phenomenon of value and its place in nature. This unique approach to the alleged conflict between science and religion can add a more sophisticated philosophical dimension to our popular discussions, which often reproduce their basic worries at too superficial a level.

My goal in the rest of this introduction is to describe some of the historical and philosophical context needed to understand the emergence and reception of Nietzsche's Darwinian Life-philosophy and the criticisms it encountered. First, I sketch briefly the historical context of late nineteenth-century German thought that informed these four thinkers. Next, I discuss how Nietzsche and his critics thought about the core values embedded in scientific investigation and religious practice, so that we can appreciate the special concerns that these raised for evolutionary thinking. The title of this book refers somewhat provocatively to Nietzsche's Darwinian "religion," and in this section I explain why understanding Nietzsche's appropriation of Darwin as a religious project is apt. I conclude by discussing a central issue inherited from Kant that permeated these discussions of the implications of life for human values. This was the issue of biological *teleology*, or the ways in which understanding living things required concepts of purpose, goal, and drive—an essential striving *for* something—in a way that understanding inanimate things did not. In each chapter, I analyze how Nietzsche and his critics understood the teleological character of life and show how this determined their approach to the relevance of evolutionary biology for philosophy and religion.

Science and Religion in Nineteenth-Century Germany

Let us begin with part of the historical background to these debates, in admittedly broad strokes. The philosophical landscape of continental Europe at the turn of the twentieth century can initially be approached through the lengthening list of competing "-isms," some new and some old, that battled over common theoretical territory. Materialism, pragmatism, positivism, idealism, vitalism, psychologism, irrationalism, logicism, and historicism each had a stake in the question of what normative worldviews (*Weltanschauungen*) were rationally sustainable, given the advancing natural and historical sciences and the aftermath of the Enlightenment critique of religion. Indeed, many debates in philosophy today are heirs to, and attempt to sort out, the strengths and weaknesses of these competing schools. The contours of what a *scientific* worldview is and the implications of scientific research for questions of value and meaning were common ground between these conflicting philosophical camps. Indeed, central to these debates over the implications and aims of science were questions of how to assess, and even predict, the future fate of the Christian religion in European culture. The feeling of an urgent and impending "cultural crisis" arose from a sense that all discussion of value had become rationally questionable.[15] A great gulf thus appeared to open between science and life, between knowledge of the mind-independent dynamics of nature and the need to live a life that required commitment to cultural aims, ideals, and values.

Many of the fault lines in the philosophical debates that took shape around these issues in late nineteenth-century Germany appeared as the reign of the great synthesis of Hegel's idealism came to an end and the empirical sciences were no longer explicable in terms of a comprehensive and foundational philosophical system. Hegel's speculative synthesis of mind, logic, social and cultural development, the structures of intelligible reality, and the secularized Christian values of bourgeois European culture no longer seemed plausible.[16] The lack of a comprehensive philosophical synthesis and the flourishing of new, specialized scientific disciplines generated worries over how to unify the picture of nature emerging from the sciences and how to evaluate its implications for philosophy. An example of this was the question of the difference between the natural sciences and the human sciences posed in the famous *Methodenstreit* (method-controversy) in Germany between the *Geistes-* and *Naturwissenschaften* that happened in the final decades of the nineteenth century. Heinrich Rickert is mostly known in the English-speaking world through his contribution to this *Methodenstreit*, in which he defended a thesis of "disunity" (taken from his teacher Wilhelm Windelband) between *nomological* sciences of nature aimed at the discovery of general laws and *ideographic* sciences, such as history, that aimed at examining particular, unrepeatable events and individuals.[17] These different sciences, Windelband and Rickert argued, had different explanatory aims, and these conflicting *goals*, not the different topics or "stuff" in the world that constituted their subject matters, gave rise to the incommensurability of their methods, the disunity of their rational structure, and the fragmented character of the worldview that was generated by them.

The rise of critical history and the German historicist tradition, and later the revolution of Darwin and the life sciences, all appeared to pose additional challenges to sustaining the sense that humans could confidently orient their individual and collective lives in relation to secure ideals and values.[18] By rejecting overarching metaphysical systems and rendering ethical and religious ideals questionable, these circumstances changed the status of philosophy with regard to the sciences into a problem that required justification. Philosophy appeared to be at risk of suffering the same fate in academic and intellectual life as its erstwhile companion, theology. Nietzsche's demand from his book *Beyond Good and Evil*, which is analyzed in the following chapter, that "psychology be recognized once again as the queen of the sciences, which the other sciences exist to serve and anticipate" rhetorically performs the *coup d'etat* that threatened philosophy at this time.[19] Of course, we must ask what kind of philosophy was threatened in this way, if any, as well as what kind of philosophy might rise to the challenge, and this is one of the core issues debated by Nietzsche and his critics.

As a consequence of the fragmentation of the sciences and the challenge to philosophy, a host of traditional problems and aspirations—those of value, ethics, meaning, and even the nature of the cosmos as a unified whole—were left in a precarious position. Such traditional objects of philosophical and religious inquiry, and even striving, seemed to misplaced next to the most rigorous modes of inquiry into the world, which assumed a certain indifference to questions of value in favor of a focus on causation and mechanism. This situation only became more complex as new, specialized disciplines of experimental psychology, sociology, and anthropology emerged that once again shattered any hopes for an easily won conception of humanity's place in the cosmos as a whole, along with hope for the unity of the sciences, which appeared to be the only means of rationally attaining it.

These problems and the solutions to them that were debated during this immensely creative and fervent period are still with us. So, too, is the impact of social and political developments that spread eighteenth-century Enlightenment ideals of rationality and autonomy through society at large. The Enlightenment critique of superstition and miracles—and of the rational foundation of religious belief—was radicalized in the nineteenth century and went on to spread through popular as well as elite and highly educated cultural circles. Secularization, industrialization, nationalism, and calls for liberalization of the political institutions of European society challenged the place of religion in politics and society and changed the basic way that people encountered religion in everyday life.[20]

Furthermore, it was in this context, amidst these competing philosophical "-isms," emerging disciplines of science, and social changes, that a nonconfessional, academic, and scientific study of religion was beginning to be conceived at the turn of the century.[21] Indeed, the topic of religion floated among the disciplines of history, natural science, psychology, philosophy, and theology and their various methodological disputes—as it does today. The birth of a scientific study of religion occurred precisely in this philosophical context, in which the very nature of science and its status with regard to philosophy and theology were being heavily debated.[22] The problem of the place of value in nature and the place of values in the sciences was foundational for establishing a scientific, naturalistic, and nontheological study of religion, and it remains foundational for the field today.

The figures in this study attest to the fervent nature of debates over religion and science in this context, and these four thinkers must be understood against the broad background of these concerns. Nietzsche was a rebellious academic philologist; Overbeck, a historian of religion and theology; Simmel, a philosophical sociologist; and Rickert, an academic philosopher. These figures puzzled over the relationship between the scientific framework of their scholarly disciplines and the broader, existential questions posed by philosophical and religious traditions. Examining how figures from these different disciplinary backgrounds approached the common topic of religion and the common problem of discerning the possibility of a worldview that could meaningfully orient humanity in the modern scientific world is a uniquely valuable way to expose the underlying philosophical presuppositions of these disciplinary boundaries. As I aim to show in each chapter, these different disciplinary locations are helpful for understanding the divergent paths each thinker took to the problem of value in nature.

Life-Philosophy

The movement of *Lebensphilosophie* (Life-philosophy) that Nietzsche, Overbeck, and Simmel represent was also a product of this time. Despite clear and traceable lines of influence between Nietzsche and Overbeck and between Nietzsche and Simmel, none of these representatives of Life-philosophy called himself a "Life-philosopher" or understood himself to belong to a general philosophical movement. Rickert's 1920 book and an earlier essay in 1913 by the phenomenologist Max Scheler entitled "Attempts at a Philosophy of Life" invented the category of *Lebensphilosophie* (Life-philosophy). In both thinkers, the category captured a loose grouping rather than a more well-defined and systematic set of doctrines.[23] Scheler focused on Nietzsche, Wilhelm Dilthey (1833–1911), and the French vitalist Henri Bergson (1859–1941). Rickert's 1920 critique of Life-philosophy named these three central figures and brought Nietzsche and Simmel (and Overbeck by association) together with a much larger collection of thinkers, including Max Scheler himself, Ludwig Klages, Oswald Spengler, José Ortega y Gasset, Miguel de Unamuno, Darwin himself, Herbert Spencer, and even Oscar Wilde. Indeed, Rickert interpreted the still youthful American movement of pragmatism, as received especially through the work of William James, as another form of Life-philosophy that succumbed to the same philosophical confusions as all the others.[24] The general term *Life-philosopher* thus became both an analytic and a polemical category. But it was and remains helpful because it captures the fact that, for this otherwise wildly diverse group of thinkers, the concept of life was undeniably a central organizing principle.

Since Rickert's early study, intellectual historians have taken over the term *Lebensphilosophie* and continue to use it to group together a loose party of philosophers that flourished toward the end of the nineteenth century.[25] The "big three" are still Nietzsche, Wilhelm Dilthey (1833–1911), and the French vitalist Henri Bergson (1859–1941). Yet, as in Rickert, historians of philosophy use this term to capture a much wider group of thinkers who share very few systematic similarities outside of the use of the term *life*: Ludwig Klages,

Oswald Spengler, Friedrich Schlegel, Rudolf Eucken, José Ortega y Gasset, and Theodor Lessing. Even the neo-Aristotelian, vitalist philosopher of biology Hans Driesch has been labeled a "Life-philosopher." I use Rickert's term myself in this book, yet, as with any term that groups such a wide variety of thinkers with different intellectual genealogies, reference points, and contexts under a common heading, it is important to note that it ignores important and nuanced internal differences between them. Moreover, the use of this term is not meant to carry the polemical tones that it often does in Rickert's work. Even though Rickert viewed the category of *Lebensphilosophie* as primarily an analytical tool that was not alone meant to perform any critical labor, his stinging polemic, unfortunately, still colors the use of this term today.

I use Rickert's term *Life-philosophy* as a description of Nietzsche, Overbeck, and Simmel because it astutely captures the centrality of the concept of life in their reception of biology, Darwin, and the mediation of religion and science. The thinkers in this study, I argue, form a distinct "Nietzschean" strand of Life-philosophy because of their focus on the problem of value centered on questions of human agency within the teleological, goal-directed character of biological life. Moreover, I have selected these figures because they capture core systematic points of disagreement between biological and rationalist, or transcendental idealist, perspectives on the problem of value, and so their disagreements apply to the mediation of biology, religion, and science in the other *Lebensphilosophen* as well.

The criterion of selection used to identify this Nietzschean strand is thus both genealogical and philosophical. Isolating these thinkers allows this study to go more deeply into the foundational assumptions that were at issue in various responses to the significance of Darwin and biology for philosophy and religion. Unlike many of the other vitalist thinkers at the turn of the century, the Nietzscheans applied their reflection on agency and value in nature to the mediation between religion and science, and it was this strand in particular that Rickert targeted as the most sophisticated form of biological philosophy. The systematic way that the Nietzschean Life-philosophers understood the problem of science and life distinguishes them from the broader collection of thinkers who have been understood to be *Lebensphilosophen* in Rickert's study and in contemporary studies of the variety of nineteenth-century vitalist and biological philosophers.

This specifically Nietzschean strand of *Lebensphilosophie* resulted from a unique blend of inherited ideas, from German Idealist emphasis on the autonomy of normativity and the transcendental character of rationality to criticisms of Enlightenment rationalism in F. H. Jacobi and Schopenhauer and to Spinozist pantheism and the tradition of German *Naturphilosophie* in Goethe. Life-philosophy emerged out of a unique confrontation between these philosophical currents and reflection on what the sciences had to say about the diversity of life in its many forms, and this inheritance shaped the philosophical reception of the Darwinian revolution in evolutionary biology.[26] The "philosopher of faith," Friedrich Heinrich Jacobi (1743–1819), is important for understanding the genealogy of this philosophical constellation because of his formulation of the problem of nihilism as the basis of a critique of Kant and a general challenge to Enlightenment rationalism.[27] Jacobi appropriated Hume's naturalistic and skeptical critique of reason, claiming that a nonrational "faith" is the basis of reason, and thereby attacked the rationalist claim that reason could be self-grounding. But Jacobi's "faith" was not the natural passions and affects of Hume, but a distinctly Christian religious faith (*Glaube*). Against the rationalist Kantian tradition, and against his Humean influence, Jacobi argued that Christian religious faith (*Glaube*), which was independent of and prior to knowledge (*Wissen*), was the only way to avoid a nihilism that ensued through an exclusive, and excessive, commitment to reason and rationality.[28] The critique of rationalism is a significant theme in Nietzschean Life-philosophy, which, while very different from Jacobi's turn to traditional Christian faith as its solution, is also rightly interpreted as a challenge to the claim of the autonomy of reason so important to the classical tradition of German Idealism in Kant, Fichte, Schelling, and Hegel.[29] Nietzsche's Life-philosophy also takes up Jacobi's problem of nihilism while giving it a decisively naturalistic, biological diagnosis that uses the resources of his unique appropriation of Darwin and biology to overcome this troubled modern condition.

The more decisive precursor to Life-philosophy, even deemed by Rickert as an arch Life-philosopher himself, was Arthur Schopenhauer. Schopenhauer's critique of Hegel's rationalism, his famously pessimistic conception of desire, and his metaphysics of the supra-rational will as the driving force behind the phenomenal world and self-conscious agency, all announced themes that would later become central for Nietzsche, Overbeck, and Simmel. Schopenhauer set the agenda through his claim that the basic and most urgent problem for philosophy concerns the validity and source of judgments about the value of life itself. Schopenhauer inspired the Life-philosophers' criticism of merely abstract, academic, and scientific philosophy, as opposed to a philosophy that grows out of the basic drives of life and responds to the existentially urgent metaphysical need that was at the basis of religions. Indeed, Schopenhauer too was in conversation with the life sciences. He endorsed the views of an important "transformationist" precursor to Darwin, Jean-Baptiste Lamarck, who still haunts discussions of inheritance in biology.[30] In dialogue with Lamarck, Schopenhauer came to view living things as manifestations of a primal "will to live," a fundamental striving that brings life into being, shapes it, and maintains it. The Schopenhaurian concern with the "will to live" and the problem of the value of life became crucial to how the Life-philosophers understood religions. Nietzsche, Overbeck, and Simmel were all readers of Schopenhauer, and his influence will be evident in the chapters on each thinker.

Despite their nuanced disagreements, Nietzsche, Overbeck, Simmel, and Rickert share a set of assumptions about value that can be further clarified against a broader backdrop of nineteenth-century philosophy as a whole. As Hans Joas argues, Nietzsche's departure from the classical tradition of Kant and German Idealism is signaled in the very use of the concept of value, which came to occupy a central position in philosophical discourse about ethical and meta-ethical questions. He writes, "The philosophy of value has its genesis at precisely that point where faith is lost in the historicizing variants of a way of thinking that asserts the identity of the true and the good."[31] Joas agrees with the historian of late nineteenth-century philosophy Herbert Schnädelbach in seeing the central philosophical categories of *value* and *validity* that rose to prominence in figures like Nietzsche and the important forerunner to Baden neo-Kantianism, Hermann Lotze (1817–1881), as modern variants of the ancient ethical category of the good.[32] For Schnädelbach as well, the new use of the terms *value* and *validity* in philosophy signaled new meta-ethical problems that arose because the ontological and rational status of claims about what the "good" *for* humans and *of* humans was could no longer be taken for granted. All of the thinkers in this study share a post-Kantian preoccupation with accounting for the genesis and validity of value-claims—including Schopenhaurian claims about the *value of life*—in relation to scientific research on the nature of living processes.

Hermann Lotze was the forefather of Baden neo-Kantianism. He and Nietzsche thus formed the starting points for two diverging genealogies in the philosophy of normativity and value that engaged with biology.[33] The Nietzschean and antirationalist strand focused on Darwin and culminated in the work of Georg Simmel, while the Southwest (Baden) school of neo-Kantianism that started with Lotze's student Wilhelm Windelband (1848–1915) culminated in the strict rationalism of Windelband's star student, Heinrich Rickert. By focusing on Nietzsche's Darwinian approach to religion and presenting critics of it from various disciplines, this study isolates these two representative and competing approaches to the relationship between philosophy, science, and religion in turn-of-the-century philosophy.

One final historical note must be made before discussing the main philosophical questions at the heart of the disagreements between Nietzsche and his critics. The decline of Hegel's idealism and all-encompassing philosophy of *Geist* cleared the ground for the two diverging paths in the theory of value taken by the Life-philosophers and the neo-Kantians. The break between Life-philosophy and classical German Idealism is immediately apparent in the Life-philosophers' rejection of Hegel's philosophy of religion.[34] Nietzsche, Overbeck, and Simmel dismissed any purely harmonious reconciliation between faith (*Glauben*) and knowledge (*Wissen*), between the motivating aims of religious desire and the scientific and philosophical aims to arrive at a rational comprehension of reality as a unified, coherent whole. A comprehensive system of logic that would ground the empirical sciences, a system of thought's own comprehension of its own free self-development as the condition of the possibility of knowledge, could no longer be seen to culminate in what might in any degree, for either right or left Hegelians, be understood as "knowledge of God" or "the Absolute" or to satisfy the longing for such knowledge. The thinkers in this study were thus charged with the enormous task of understanding the relationship between the ideal and the real in the aftermath of Kant and Hegel and amidst the flourishing of both the natural and historical sciences.

Values in Science and Religion

Nietzsche, Overbeck, Simmel, and Rickert all approached the relationship between religion and science through the question of the place of value in nature. But how do these two topics relate to one another? The answer to this question lies in a set of claims accepted by the thinkers in this study that shaped their approach to values and generated the puzzle of their place in nature in the first place. The first of these claims is that all human action, including the activity of thinking, is *teleological*; it is directed by and aimed at ideals, values, and norms that guide and are meant to be realized in or through action. The second claim is that some of these values have the peculiar characteristic that they are taken to be *normative* and are understood to be *intrinsically* valuable. That is, they are taken to carry special authority and weight and are pursued *for their own sake* rather than instrumentally as means for the sake of some other value. On the basis of these general claims about what values were and how human agency was oriented by them, these thinkers claimed that the various spheres of European society were constituted by the ideals they offered to orient the teleological structure of human agency. The spheres of science, religion, politics, ethics, art, and even education were all constituted by the unique values that they aimed at, and the guiding values in these domains all bore just these properties of being "intrinsic," "normative," and "non-instrumental." These spheres gave us the ends to which the resources of material life, and biological life too, might be devoted as means.

The activity of scientific investigation can be taken as a concrete example to illustrate this approach. All thinkers in this study understood the very idea of science to be the pursuit of the non-instrumental value of knowledge, of an accurate understanding of the world *for its own sake*. Recognizing this aim of science was crucial for identifying just what science is, and it was non-instrumental in the sense that science was not pursued for any other purpose than the understanding that was achieved by it. The fact that science was thought of as an activity aimed at a value, one that contained normative principles that distinguished *valid* from *invalid* thought, meant that it too could only be made sense of within the sphere of human agency. In Rickert's own words in the epigraph to this introduction, "The idea of a science is always the concept of a task to be carried out."[35] This task was that of understanding nature *rightly*, independent of the utility of this knowledge and independent of its implications for the values in other domains—aesthetic, ethical, and religious. Of course, this understanding of science as guided by a normative value made the notion of value-free or value-neutral science highly problematic, and this is a crucial point for the philosophy of science that came out of this tradition. There was something important that the characterization of science as value-free captured, but this description was highly misleading. The right way of expressing the relation between science and values was that the value of getting the world right *ought* to be pursued regardless of its implications for other cultural values. Value-freedom *from non-epistemic values* was, ironically, itself a value and an ideal, as was the value of understanding nature that science hoped to achieve.

The problem that these thinkers faced in Darwinism and the evolutionary understanding of nature can now be stated more precisely. The problem was how these so-called "intrinsic," "non-instrumental" values could be made sense of within the explanatory frameworks of the biological sciences and the pictures of natural processes emerging from them. Could such values be understood as natural, or as having their source in nature? The puzzle of discovering the origins of such cultural values arose out of the peculiar properties of these values, which seemed to be autonomous with regard to the strictly biological needs that humans shared with other living things. This was a question of the natural sources of values, but it was thereby also a question of the origins of reason and rational agency that was aimed at these now peculiar sorts of goals—namely, the intrinsic, non-instrumental values mentioned above.

This question of *origins* was only one aspect of the problem. The next and equally crucial problem was how to assess what the potential evolutionary and biological origins of values meant for the special, normative status—the *validity* (*Geltung*), as these philosophers called it—that certain values were thought to possess. For instance, it seemed that the value of truth *for its own sake*—the constitutive value of science—was special because accepting it or rejecting it could not be considered a matter of mere individual preference. A rejection of the value of truth appears to be rationally defective in some way. This was a core problem upon which the Life-philosophers and Rickert disagreed. Whereas the Life-philosophers wanted to ground this special normative "seeming" in natural processes that permeated all forms of life—human and nonhuman—Rickert thought that this could not be done without relinquishing this special status altogether. The problem of normativity in this debate was posed as a question of whether any ends could be vindicated as *valid* against the picture of humanity presented in Darwinian evolution.

While the origin of the central value of science—knowledge *for its own sake*—posed a peculiar problem, similar problems were raised by the phenomenon of religion. Religious life too, these thinkers proposed, could only be understood in the context of the goal-directed, teleological structure of all human action that was aimed at intrinsic values. Religions, like the sciences, also aimed at realizing core values, and these values were in part constitutive of the kinds of human practice that religions were. While the idea that science is driven by a value may have been more controversial, there is less difficulty in noticing that religions involve claims about what matters, about which kinds of life that one might lead are worthy to pursue, and what ideals might lead to forms of human satisfaction and fulfillment. Yet, while religious values are often understood as simply equivalent to moral values regarding how to treat other humans (and animals), these thinkers recognized a difference.

Nietzsche and his critics thought that religious values possessed a unique character that made them especially philosophically interesting. To capture this, it will help to draw on a recent and insightful essay on religion by the American philosopher Thomas Nagel. Nagel writes that the "religious temperament" seeks "a view of the world that can play a certain role in the inner life," and that religions "supply an answer to the question of how a human individual can live in harmony with the universe."[36] As Nagel's description helps bring into view, religions are views of reality that have practical consequences for questions of what matters and why, and of how one "ought" to orient oneself in light of the human condition, which is in part determined by biological factors like mortality. For this reason, religions stand in the starkest opposition to value-free knowledge; indeed, they fundamentally seek knowledge *of* value, or knowledge that will contribute to the realization of value. The vocabulary of "transformation," "salvation," "enlightenment," "virtue," and so forth that populates religious traditions indicates that knowledge is meant to serve the purpose of achieving certain kinds of human goods. The theologian Schubert Ogden also helps to get at this point through a phrase that recurs throughout his work in the theory of religion. Ogden writes that religions are those cultural forms that ask and answer what he calls the "existential question": namely, "What is the meaning of ultimate reality *for us*?" Ogden points out that this central, "existential," question is unique because it couples the metaphysical and the moral; it presupposes that reality has implications for human values. Different ways of life are assessed as being more or less *in line* with reality; therefore, for the "religious temperament," the real itself is sought out as a criterion of value.[37]

These definitions of religion are philosophically important because they show the constitutive religious question to be entangled with the aims of philosophy and science. The "religious temperament," as Nagel characterizes it, envisions reality such that there can be a way of life that is in line with it. Religions thus possess an "extra-scientific" *interest* in how we understand and come to know the natural world, and this *interest* is well captured by Ogden's term "existential." Knowledge is not simply to be sought *for its own sake*, but for the sake of our existential interest in discovering a way of living that brings humans into harmony with their nature, and with the nature of the world as a whole.

Another definition offered by the philosopher of religion Kevin Schilbrack brings this out as well and takes this point a bit further. Schilbrack writes that "religions are composed of those social practices authorized by reference to a super-empirical reality, that is, a reference to the character of the Gods, the will of the Supreme Being, the metaphysical nature of things, or the like. In short, I define religion as forms of life predicated upon the reality of the super-empirical."[38] Schilbrack helpfully highlights an obvious fact that is nonetheless often missed: the very aim of religion involves it in metaphysical and meta-ethical inquiry. Religious worldviews are comprised of both practical *forms of life* and theoretical understandings of what the world is like that *justify* these forms of life. This definition points to a feature of religious worldviews that may be obvious, but whose philosophical implications are rarely thought through fully: whether ultimate reality is viewed as God, Nothingness, Brahman, a pantheistic deification of nature, a teleological nature, a mechanical nature-system, or as the Form of the Good matters to our ideas of what kinds of human actions and dispositions can be conceived to be "in harmony" with it. The Life-philosophers and Rickert brought attention to the fact that the value that religions seek to realize was precisely the value of a way of life in harmony with world as a whole. The final argument of this book is that, for these thinkers and others in the wider tradition of German thought which they carried forward, religion became a topic of significant philosophical interest precisely because it sat curiously at the nexus between two worlds, the world of the "is" and the world of the "ought," and it promised the possibility of binding these two worlds together.

How do these conceptions of science and religion help us to understand the reception of Darwinian naturalism by philosophers in Germany? This question has two of the same components that we met when discussing science and the value of knowledge. First, both Rickert and the Life-philosophers were interested in understanding whether the religious value of acting in harmony with reality and the scientific value of pursuing knowledge *for its own sake* can be understood to arise out of biological needs and interests, especially ones that humans share in common with other living things. This was the *genealogical* question of the origins of these values. While perhaps not maladaptive, in the Darwinian sense, such values at least seemed to leave behind the logic of adaptive benefit that governed most biological thinking about the aims of organisms in the struggle for life generally.[39]

The second question raised was how one might act in harmony with reality when this reality is now understood through the dynamics of evolution and the features of the process that generates life's many forms. For humans (and perhaps other animals), life did not just happen on its own; it had to be lived, and lived self-consciously through an orientation to values. It was just here, on this complex but hugely decisive point, that the problem of value in Life-philosophy and neo-Kantian philosophy was situated. Darwin convincingly argued that life forms arise and go extinct over the *longue durée* of geological time through struggle and competition. What forms of agency did humans have within this process? What forms of fulfillment were possible? What was the moral meaning of this view of nature *for* human life? These questions, which belong to an existential hermeneutic of nature, are the questions that we encounter in the following chapters.

Values and Biological Teleology

One last topic must be mentioned to help organize the different views presented in the following chapters. The previous sections introduced the term *teleology* to capture the goal-directed character to human agency, but this term also has an important application in the biological sciences. In biology, the term *teleology* refers to the apparently purposive character of biological entities, the ways in which their organization is apparently "directed" toward some goal, as in the example of an eye being organized for the purpose of seeing, or even in terms of behaviors like hunting, foraging, or nest-building. The importance of this term for sorting out the relationship between life, value, and validity in these thinkers further indicates the post-Kantian nature of this debate. Many aspects of Kant's philosophy were important for setting the agenda of nineteenth-century German engagement between philosophy and the life sciences. These include, foremost, the distinctions between intuition and conceptual thought, between constitutive and regulative ideas, between theoretical and practical reason, between questions of fact (*quid facti*) and questions of right (*quid juris*) and, finally, focus on the normative, *a priori* presuppositions necessary for, but justified independently of, empirical investigation. However, one aspect that receives less attention than these is found in the second section of Kant's third *Critique*, *The Critique of the Power of Judgment*, which introduces his philosophy of biology. Kant writes here of the necessity of conceiving of living things as *teleological*—that is, as organized for some purpose, end, or value. Kant writes that "the structure of a bird, the hollowness of its bones, the placement of its wings for movement and of its tail for steering, etc." indicates a "unity in accordance with such a rule" of being suited for a purpose.[40] Not only does each part of an organism play a specific role in playing out its life cycle of survival, self-maintenance, and reproduction, but all the parts together seem to function toward a common, unified end. Kant's notion of teleological organization reformulates Aristotle's notion of *psuchē*, or "soul," which in Aristotle meant the organizing principle of the matter that composes a body. The *psuchē* of a living thing is the internal principle of change and motion that constitutes the living form that the material composing its body takes on.

For Aristotle, teleological concepts help us capture what sorts of things living things are—namely, purposively organized entities that strive to achieve various aims—and what properties they possess. Kant, by contrast, held that while we cannot but conceive living things using teleological and purposive language, such a principle was problematic because it could not be formulated in mechanistic terms of blind causal laws and properties of inert matter that were proper to the physical sciences. In order to resolve this problem, he argued that the teleological conception of organisms as purposively organized units must be regarded as a *regulative* idea: it is an idea that is useful for helping us investigate and discover underlying mechanisms in nature, but it is not, strictly speaking, a real, constitutive property of organisms themselves. In Nietzsche, in Darwin himself, and in discussions of Darwin and the life sciences in the late nineteenth century generally, the question of the validity of teleological conceptions of organisms resurfaced, and Kant's restriction of this concept to purely regulative status was widely transgressed. Many of the biologists Nietzsche read, and from whom he likely received his understanding of Darwinian evolution, unrepentantly relied on teleological conceptions both of organisms and of the evolutionary process generally. Even Darwin himself accepted the teleological view of organisms. Indeed, debate over the legitimacy of teleology in biology is still central to the field of philosophy of biology today.[41]

The following chapters show that in order to understand the disagreements between these thinkers about the ontology of value in nature, it is useful to examine the degree to which they made space for teleology within the Darwinian conception of evolving life. Of course, Nietzsche famously challenged one conception of teleology in the second essay of his Darwinian masterpiece, *On the Genealogy of Morality*. There, he rejected the claim that traits come to exist because of their usefulness and function and argued that the question of origins is utterly distinct from the question of function.[42] The notion of teleological explanations for the origins of traits is rejected by most contemporary neo-Darwinian thinkers, and Nietzsche can be seen to have endorsed a version of the contemporary principle that the origination of traits is random with respect to the beneficial or harmful consequences that these traits may come to have.[43] But this is not the sense of teleology that has its sources in Aristotle and Kant and that was crucial for Nietzsche's biological approach to the problem of value.

The sense of teleology that continued to be important for the negotiation between philosophy, religion, and the sciences in Nietzsche and his critics was the language of purposiveness and directedness that attends the representation of life itself. This notion of teleology represents living things as possessing some form of agency, such that it makes sense to ask what they are after, what their goal is, what needs they have, and what drives them. It explains vital function and organization according to a goal and purpose of the organism and in relation which the organism as a unified whole can be represented. The legitimacy of this language continues to fuel fascinating debates in philosophy of biology today, and for our purposes it is important to recognize that this language of agency and goal-directedness permeated nineteenth-century discussions of evolution and philosophy.

The idea that there is something that living activity is after, and that acknowledging this is important for capturing what living activity *is*, formed the basis of Nietzsche's Life-philosophy. Moreover, this notion of a basic drive that characterized life became a point of contention between Nietzsche, Overbeck, and Simmel. While differing in their notions of biological teleology that were acceptable, all three Life-philosophers challenged the autonomy and sovereignty of *Geist* (mind) that had dominated the tradition of German Idealism that they inherited, and they reconceived rationality and the distinctiveness of human agency by grounding these in drives and powers operative in the biological world as a whole. Nietzsche wanted to "translate man back into nature" and show what "Man, the animal" (*das Tier Mensch*) meant for philosophy; Overbeck wanted to distinguish truly *vital* concerns from those that were merely scientific; and Simmel claimed that his philosophy of life was to "place back into life itself everything that had been established outside of life."[44] Rickert, on the other hand, defended a fundamental separation of reason from nature—of rational agency guided by *valid* values from any *telos* that appeared at the level of "mere" biological life. The question of what relation human agency bore to the rest of the living world brought the life sciences into contact with the fundamental problems of Western philosophy and religious thought. The following chapters recount a complex and rich set of questions that arise when one turns to science and to evolution for answers to the existential puzzles of the human condition.

1

Friedrich Nietzsche: A Darwinian Religion

If one statement from Nietzsche's vast corpus could be selected to capture the sense of validation he found in Darwin's theory of evolution for his own project of a "revaluation of values," it would be this: "*Life is something essentially non-moral.*"[1] This phrase appeared in the preface that Nietzsche added in 1886 to his earliest published work from 1872, *The Birth of Tragedy*, and it signaled Nietzsche's sense of the continuity between his concerns as a young professor of classics in Basel and his mature views as the nomadic iconoclast of Western philosophy. It also captures a puzzle that surrounded the philosophical implications of evolutionary biology in the strand of German thought that Nietzsche came to embody. For Nietzsche in particular, there were three main questions that Darwin's discovery raised: First, if life is non-moral, then how could moral values have arisen in the first place? Second, if moral values do not capture the dynamics of the living world, then what values —if any—do? Lastly, what would it take to *affirm* such a world? These clearly philosophical, and even existential, questions were raised by the evolutionary ideas floating around the sciences of Nietzsche's time. Moreover, I argue in this chapter that these were the questions that became essential as Nietzsche and his critics sorted out the relationship between religion and science.

Some disclaimers must inevitably be made before plunging into Nietzsche's thought. Nietzsche perhaps more than any other thinker in the history of philosophy is a figure with many faces, and the task of "getting Nietzsche right" is one that is fraught with difficulty. It is obligatory before beginning any discussion of what Nietzsche thought to note that attributing to him any sustained position in ethics, political philosophy, philosophy of science, or philosophy of religion is a nearly impossible interpretive task. Here, too, we must admit that the following interpretation will inevitably meet strong objections based on counter-evidence from Nietzsche's own vast *corpus*. Part of the difficulty is due to Nietzsche himself; he does not often provide explicitly formulated arguments for his positions in a form that contemporary philosophical readers have come to expect and demand in philosophical work.[2] He also often makes claims in various books that are undeniably in tension with one another. Nietzsche certainly did not compose his texts for readers who approach his work as skeptics looking to be convinced through rational argument alone—indeed, the reason for this is rooted in his understanding of human psychology and the nature of reason in general, which I discuss in this chapter.[3]

Another problem is simply the breadth of topics in Nietzsche's texts, to which any simple reduction to a basic or privileged concern does not do justice. Nietzsche wrote quite a lot over many years in his published works, unpublished notebooks, and extensive letters with philosophical content. Although I argue that a lasting concern or guiding thread can be traced from Nietzsche's earliest writings to his final ones, this should not camouflage the stunning variety of individual aims, argumentative strategies, brilliant flashes, and stylistic forms both within and among his various books at different stages of his life. In fact, the guiding thread that I concentrate on is one that is almost constructed by Nietzsche himself as he turned, in the late 1880s, to reflect on his early works and to discover in them the seeds of his later concerns, especially on the topics of science, life, value, and Christianity. I rely on these late, retrospective readings but am aware of the growing body of philological research that has challenged claims of continuity and the attempt to find an essential Nietzsche.[4] Rather than argue that Nietzsche's total literary output contained only one definitive account of the relationship between religion and science, one concept of life, or one theory of what Christianity is, I instead piece together a coherent, recurring, and fundamental set of issues that shaped the reception of Nietzsche by his early critics and, I would also venture to claim, is the source of much contemporary constructive interest in his writings.

While simplifying things greatly, one can highlight two basic camps today that claim Nietzsche as their intellectual inspiration. One of these is the Foucauldian, postmodern, and postcolonial tradition, which in contemporary anthropology and cultural studies has taken the concept of power to be fundamental for social theory and has aimed to analyze often hidden power dynamics involved in shaping cultural identities and values—religious, political, or otherwise.[5] In the process of refining Nietzsche's method of genealogy and uncovering dynamics of power in processes of enculturation, belief formation, and identity formation, this tradition has cast into deep suspicion the entire concept of normativity and validity, and has even gone as far as to question the legitimacy of ethics as a philosophical enterprise. For such critical theorists, Nietzsche helped us to see through the pretenses involved in the privileged values of truth and knowledge that had guided his philosophical predecessors, and he revealed the exercise of subtle forms of social domination behind the search for ultimate values and sources of value. Nietzsche's invention of the genealogical method of analysis and critique suited the purpose of this unmasking of power relations, forms of dominance, and even desire underneath the mantle of sober rationalism and taken-for-granted normative values. But it left the question of what the human good is in its wake.

Historians of philosophy in the Anglo-American and German traditions have revealed a rather different Nietzsche. A growing body of evidence shows Nietzsche's deep engagement with and defense of science, and of the life sciences in particular.[6] Rather than reject science and deconstruct notions of truth and knowledge, many interpreters show Nietzsche to be a staunch defender of science and even a metaphysician who saw his metaphysics confirmed by the sciences.[7] This Nietzsche did not reject normative values; he sought new and better ones that were defensible on the basis of rigorous scientific inquiry. Many interpreters today rest the contemporary relevance of Nietzsche in his naturalistic account of the origins of human values, which was inspired and validated by the scientific thought of his day. This Nietzsche passionately pursued a new and better notion of what the human good is.

The goal of this chapter is not to debate the strengths and weaknesses of these different readings, but to explore Nietzsche's work in a way that will expose a key ambiguity that is at the heart of their divergence. I aim to show that these two paths diverge in response to an ambiguity in how Nietzsche related the realm of life to the realm of value, and that this ambiguity was also crucial for early critical responses to his work. Analyzing Nietzsche's notion of life is crucial for the later chapters because of the confusion over how to combine Nietzsche's project of "revaluing values" and combating nihilism with his appropriation of biological and scientific thought that his early critics—Overbeck, Simmel, and Rickert—all seized upon. Their conceptions of life, and of the significance of the concept of life for understanding the relationship between science and religion, were responses to Nietzsche's unique combination of naturalistic investigation and normative, ethical, critique. Moreover, analyzing Nietzsche's concept of life allows us to understand his reception of Darwin. The general problem of the relationship between life and value became the point of confrontation between philosophy and the natural sciences, and it was on this ground that these thinkers debated the philosophical implications of evolution.

Whether Nietzsche was an antinormative postmodern, or a defender of science, a metaphysician, and even a moralist, his attempt to understand ethical and religious values in the context of the living world of organisms attempting to flourish, and to secure the means of flourishing in nature, is still the reason why many have returned to him as a resource for contemporary ethical theory. But there is another side to Nietzsche the *Lebensphilosoph* that is less emphasized by contemporary retrievals of his work for ethical theory or cultural anthropology, and that his early critics saw as fundamental. To use Thomas Nagel's phrase quoted in the introduction, Nietzsche the Life-philosopher had a "religious temperament"; he wanted to discover a new way of life that was aligned with the way things are. This Nietzsche seized upon the evolutionary view of nature in such a way that it could come to "play a role in inner life" and because it helped humans understand what it meant to flourish as a creature of nature. Understanding human moral life in terms of flourishing as a natural creature, a product of a natural history of evolution, could allow Nietzsche to develop what I call a "realistic idealism." That is, it could help inform the religious quest for a way of life that would lead to ultimate fulfillment by telling us something about the nature of the world in which this fulfillment was sought. The following chapter aims to show that Nietzsche's concept of life did not erase value from nature but showed that valuing and life were fundamentally entwined to the extent that life was both the ultimate source and criterion of value.

Nietzsche's Concept of Life

Max Scheler and Heinrich Rickert early on recognized the centrality of the concept of life for Nietzsche, and they enshrined this by designating him as a *Lebensphilosoph*. It was also recognized as central in Martin Heidegger's highly influential Nietzsche lectures in the late 1930s through the 1940s. However, Heidegger's writings on Nietzsche subordinated the term *life* to its synonym, the *will to power*, thus partly downplaying the biological resonance of this fundamental notion. Moreover, Heidegger's influential interpretation of Nietzsche led the excerpts from Nietzsche's notebooks published posthumously in 1906 as *The Will to Power* to become a privileged text. This text was misleadingly publicized by Nietzsche's infamous sister, Elisabeth Förster Nietzsche, and his friend Heinrich Köselitz as the culminating *magnum opus* that Nietzsche had been planning before his breakdown in 1889. As a result of Heidegger and this unauthorized posthumous publication, the "will to power" gained priority over the notion of life. Heidegger explicitly wrote off the naturalistic, biological, and Darwinian reading of Nietzsche as superficial and declared that the unpublished material in the posthumous *The Will to Power* gave more insight into the meaning of Nietzsche's philosophy than his published works. The influence of Heidegger's critique of Nietzsche's "will to power" as a metaphysical doctrine—meaning one that attempted to characterize "the being of beings"—and his subordination of the significance of Nietzsche's published works to his unpublished notes prevented Nietzsche's deep engagement with the biology of his day from attracting much philosophical interest.

Recent historical and philosophical work on Nietzsche has challenged Heidegger's dismissal of biological themes in Nietzsche's published texts. While it is still questionable whether or not Nietzsche read Darwin himself, he voraciously read German biologists who critically appropriated Darwin, such as William Rolph, Wilhelm Roux, and Carl von Nägeli, and he followed discussions of Darwin circulating in his day.[8] Nietzsche's engagement with biology stemmed from a core conviction that to understand human ethical and religious values, and even to critique and evaluate them, one had to understand them through the characteristics of living organisms more generally. And this required investigating the *teleological* concepts prevalent in the life sciences. These were concepts that captured living entities in terms of what they were after—what they wanted and needed and what it meant for them to do well or badly. Nietzsche turned to biology because he thought it was fundamental for providing the context within which human ethical and religious values would be shown to fit into the natural world.

As it divided Nietzsche's earliest interpreters, the problem of interpreting Nietzsche's twin notions of "life" and "will to power" as metaphysical, biological, psychological, or even phenomenological persists in contemporary debate. As ever-present as the concept of life seems to be in Nietzsche's books and notes, it also bears a strongly metaphorical quality, and Nietzsche's unsystematic style of writing notoriously makes philosophical analysis difficult. Even though the concept of life is ubiquitous, it is rarely given explicit definition in the variety of contexts in which it is used. Despite these interpretive difficulties, one can scarcely pick up a work by Nietzsche without reading castigations of those cultural forces that "deny life" and those that can only "affirm" it by falsifying it or looking beyond it. When he does identify characteristics of life explicitly, they are often quite general and vague features such as "strength," "growth," and "ascension"—or their opposites, "weakness," "decline," "degeneration."[9] It is clear that the concept of life poses strong challenges for interpreting Nietzsche's philosophy, yet it is so central that it cannot be avoided.

The concept of life first figures prominently as a term of art in the title of the second essay of Nietzsche's early *Untimely Meditations* (*Unzeitgemäße Betrachtungen*), "On the Use and Disadvantage of History for Life" (1874), and it is in this essay that Nietzsche gives the earliest direct definition of life in his published work. There, Nietzsche describes life in richly metaphorical language as "that dark, driving, insatiable power craving itself."[10] While this early definition is highly figurative, it nonetheless contains crucial meanings that persist in Nietzsche's later use of the term and ought not to be written off as mere metaphor. Here, life is an erotic notion—an appetite, an animating "drive," a "power," a desire and hunger. But just as importantly, Nietzsche names an *object* and *aim* of that desire and thirst—namely, life itself. Life here is an appetite and a drive that is naturally directed at itself and so in this sense is self-valuing, it wants more of itself and seeks itself as its own *telos* and as that which will satisfy its own "craving."

Nietzsche's goal in the second *Untimely Meditation* is to evaluate the discipline of history in terms of its service to this drive—a drive that Nietzsche finds to be at the origin of great cultural accomplishments, virtuous and noble individuals, and at the origins of phenomena like religions and even science itself. Nietzsche's early meditation on the relationship between science and life was precipitated by the problems of historical consciousness that were already beginning to occupy many other thinkers of his generation, including his friend Franz Overbeck and later Simmel and Rickert.[11] The central question that precipitated the so-called "crisis" of historicism was similar to the existential question posed to biology outlined in the introduction: How can historical knowledge be useful for living a life in the present, for determining what to value and pursue? Can history have meaning *for living*, or did it merely undermine a naive self-confidence in ideals by showing us what fate they had met in the past or that ours are simply a few of many ideals that have animated human lives across history? In answering these questions, Nietzsche's early essay already rejected a positivist, value-free conception of science that stripped it of ethical content and import. Nietzsche's attack of history was an attack of the positivist pursuit of knowledge *for the sake of knowledge*, disconnected from the drive of life to seek after and "crave" itself. It was precisely the separation of science from this erotic source of value that threw into question its existential and ethical utility. The remedy for this situation, Nietzsche claimed, was to recognize that life and its striving to realize ideals was present in science itself, in the form of the very ideals of knowledge and intellectual understanding that were its objects of desire.

In this early essay, life already plays the controversial dual role that it continues to have throughout Nietzsche's work. *Life* is both a descriptive term for a fundamental drive that is manifest in human action, thought, and desire, and a normative criterion of value. Service to the drive of life, to its desire for itself, acts as a criterion against which other values—in this case the value of historical *Wissenschaft* as knowledge of the past *for its own sake*—must be ranked. In this essay, Nietzsche pits the *telos* of value-free knowledge against the *telos* of the vital impulse for more and greater life. The difference between the end pursued by science—knowledge for the sake of knowledge—and the end pursued by the appetite that *is* life leads Nietzsche to question the motivations that led his contemporaries to pursue ever more detailed and rigorous knowledge of the past. Indeed, the diagnosis is even stronger: Nietzsche blames the value-free aim of science for contributing to the enervation of an instinctive, vital interest in creating and realizing lofty ideals, and this was the root of an existential crisis of European culture.

Nietzsche's second *Untimely Meditation* also offers initial traces of a theory of religion: "A religion, for example, which for the sake of pure justice is to be transformed into historical knowledge, a religion, which is to be known scientifically through and through, has at the end of this at the same time been destroyed. The reason for this is that in historical accounting so much that is false, crude, inhuman, absurd, violent comes to light that the mood of pious illusion in which alone anything that wants to live can live necessarily shatters."[12]

Here, the standpoint of life is contrasted with a disenchanting pursuit of scientific knowledge and learning. The "pious illusion" that is required in order to maintain a life is unable to withstand the sense of contingency, even violence, that makes the confident pursuit of lofty ideals appear naive and futile. In this passage, Nietzsche offers religions as concrete sites at which the conflict between science and life plays itself out; religions as objects of scientific knowledge and religions as living pursuits are fundamentally at odds. Religions require a kind of idealism, not in the epistemological sense that the objects of experience are constituted by the activity of the mind, but rather in the sense of the attempt to realize core values in a world whose malleability and receptivity to them is unclear. Indeed, values were to be realized in a world that may even be hostile to them, and this was only to be discovered in the course of making the grand experiment of living them out. This requires, as Nietzsche wrote, a degree of "pious illusion," an unreflective confidence and orientation toward the present and future, that history could only disenchant, and from which science could only be a distraction.

In the guise of this meditation on the value of historical knowledge, Nietzsche's second *Untimely Meditation* took over a fundamental issue that he had already raised in his first published work, *The Birth of Tragedy*, in 1872. Nietzsche called this the problem of the *value* of truth. The term *life* as an organizing principle did not make it into *The Birth of Tragedy* until Nietzsche appended his retrospective "Attempt at Self-Criticism" later in 1886. In this late preface to his early work, Nietzsche wrote that life, and the so-called "problem" of science and life, was at the very center of his early investigation of Attic drama. He famously declared in the preface, "And science itself, our science—yes, what does all science mean as viewed as a symptom of life? For what, or even worse, from where does all science come? How?" The next section elaborates: "What I then came to comprehend, something terrible and dangerous, a problem with horns, not necessarily a bull, but anyway a new problem: today I would say that it was the problem of science itself."[13] Nietzsche retrospectively claimed here that the central problem of *The Birth of Tragedy* was the problem of the relationship between science and life. And this problem was, more specifically, how to account for the origin of science and the scientific aim of truth *for its own sake* from out of the "dark, driving power" that he originally sought to define, and with which it came into conflict, in his second *Untimely Meditation*. By famously daring "to see science under the optic of the artist, but art under the optic of life," *The Birth of Tragedy* first threw into question the *utility* of knowledge and truth for satisfying life's own deepest and most fundamental drives and needs.[14]

Nietzsche captures this problem of science and life in *The Birth of Tragedy* through the figure of Socrates, who came to personify the conflict between the practices, aims, and values of reason and knowledge, of "theoretical man," in contrast to the animating and inspiring drives that generated the tragic view of life expressed in Greek drama. Indeed, Nietzsche's first published work itself enacts a subordination of science to life by transgressing the "purely" *Wissenschaftlich* demands of his own philological discipline and subordinating them to the extra-scientific and existential aim of recovering the tragic ideal for the modern world.[15] The aim of his analysis of Greek tragedy was not truth *for its own sake*, but to recover a lost orientation to the value of life that expressed itself in, and even generated, this form of art in the ancient world. This decidedly unscientific aim, of course, led to harsh condemnations from Nietzsche's colleagues in philology and ultimately to Nietzsche's disaffection from the discipline in which he was educated. Yet, Nietzsche's early work, permeated as it was with Wagnerian ambitions and Schopenhaurian problems, indicted the conception of value-free *Wissenschaft* defended by his colleagues in philology in a way that stayed with him throughout his writings. Nietzsche's first scholarly investigation was guided by a *vital* concern with the problem of the meaning and value of life, one that looked to a forgotten and profound ideal that he found to be an animating force behind pre-Socratic Greek drama. This vital concern would continue to animate him after he left his academic position in Basel in 1879 to seek a climate more favorable to his ever-declining health and to begin his most prolific years, during which the natural sciences, and not history, became the source of his reflection on the conflict between science and life.

The Affirmation of Life

While Nietzsche's early works first introduced the concept of life and its conflict with science, his later writings further developed an account of this conflict that was informed by his turn to contemporary biology and Darwin. In order to understand this transition, I again turn to Nietzsche's late retrospective and autobiographical analyses of his early works, this time those offered in his last book, *Ecce Homo*, written in 1888 and first published posthumously in 1908. In *Ecce Homo*, Nietzsche returned to *The Birth of Tragedy* again and credited it with two crucial discoveries. The first was that it recognized the tragic art form as "the first instruction on how the Greeks got over Pessimism—with what means they overcame it."[16] Nietzsche went on to write that the goal of that early work was an analysis of the psychology of Greek tragedy as a form of art; his study of Greek tragedy was to decipher the *instincts* that produced such a form of art and the affective experience that tragedy generated in its audience. This hermeneutic strategy resembled the methodological conception of cultural studies defended by the historian Jacob Burckhardt, a colleague of Nietzsche and Overbeck in Basel. Burckhardt argued that the aim of cultural history was essentially to understand the history of states of mind and "living energies" that actively shaped human cultural life in distinct places and times.[17] Looking back, Nietzsche wrote that tragedy deserved philosophical attention because it was "proof that the Greeks were not pessimists."[18]

While Aristotle's *Poetics* argued that the affective impact of tragic drama was a purgation or cleansing of "a dangerous affect," Nietzsche described it as a way of going beyond fear and pity "in order to become the everlasting joy [*Lust*] of becoming itself, that appetite, which still also contains within itself the appetite for destruction."[19] In this passage from *Ecce Homo*, Nietzsche reuses the final aphorism from *Twilight of the Idols* to capture his earliest efforts to decipher the underlying affective psychology of tragedy. He wrote further that this affect was "the affirmation of life itself still in its strangest and hardest problems; the will to life rejoicing over its own inexhaustibility even in the very sacrifice of its highest types—that is what I called Dionysian, that is what I understood to be the bridge to the psychology of the tragic poet."[20] Following this citation from *Twilight of the Idols*, Nietzsche goes on to declare of himself that "in this sense I have the right to understand myself as the first *tragic philosopher*—that means the most extreme opposite and antipode of a pessimistic philosopher."[21]

Nietzsche's pregnant description of himself as a tragic philosopher brings the ethical dimension of his project into view. In this passage, he ties the notion of an affective, instinctual affirmation of life to his early notion of life as a self-desiring—here a self-"affirming"—appetite. This anti-Schopenhaurian, antipessimistic declaration was a plea for a confident affirmation of life over against the various historical and cultural forces that allegedly prevent, suppress, and deny it. In tragedy, the vital drive of life's self-reflexive affirmation of itself is presented as an existential solution and the source of overcoming a perceived nihilistic weariness and devaluing of life in the world that surrounded him. For Nietzsche, the vital drive of life was the source of its own affirmation, even in the face of the suffering and destruction that appeared to result inevitably from its own dynamic, creative drive.

In this set of self-reflective passages, Nietzsche describes himself in ways that frequently appear in his other texts as well: he is a "psychologist," a "doctor," and a "tragic philosopher" who wishes to overcome pessimism through an instinctive affirmation of life in the face *of what life really is*—namely, endless becoming, creation, destruction, brutal "contradiction and war."[22] The tragic worldview of the Greeks was enviable from Nietzsche's nineteenth-century perspective because it signaled to him such an instinctive affirmation, one that arose neither as the result of value-neutral investigation of the world, nor from a contemplation of the mind-independent metaphysical order of things, nor as the conclusion of a rational argument or dialectical exchange, but from the recovery of a deep instinct and hunger that was fundamental to life as a process in nature.

The second achievement of *The Birth of Tragedy*, Nietzsche claims in his retrospective reflections, was the discovery of the true nature of the figure of Socrates—or, more generally, the phenomenon of "Socratism." To Nietzsche, Socrates was to blame for the dissolution and decline of the tragic but nevertheless life-affirming worldview because he popularized a false dichotomy and cemented it in the philosophical imagination of his generation. Socrates created the opposition of rationality *against* instinct, but Nietzsche commends himself for being able to recognize this rationality as "dangerous" and "a force that saps life."[23] Through both of the achievements of *The Birth of Tragedy*—the discovery of the life-affirming instinct that produced the tragic worldview and the discovery of Socrates as the symptom and sign of its loss of vitality—Nietzsche sees himself as "high above such miserable chatter" that pits instinct against reason. He writes,

> The degenerating instinct that turns itself with subterranean vengefulness against life (—Christianity, the philosophy of Schopenhauer, in a certain sense already the philosophy of Plato, all idealism as typical forms) and a formulation of the highest affirmation born out of a fullness, an overfullness, a yes-saying without reservation, to suffering itself, to guilt itself, to everything in existence that is questionable and strange... this last, most joyful, exuberant-carefree yes to life is not only the highest insight, it is also the deepest, and most rigorously verified and sustained by truth and science.[24]

To this passage could be added many others throughout Nietzsche's corpus that describe life-denial and life-affirmation as two opposing conditions of instinct. The true underlying antagonism that creates the opposition of science and life is not rationality versus instinct, but rather the instinct to affirm life against the instinct to reject it and rebel against it. These late, retrospective passages show how Nietzsche's early "deep, dark desire that thirsts after itself" remains in his later work and becomes a fundamental concept for approaching the question of the value of life that he seeks to bring into the center of philosophical attention and to use as a diagnosis for the nihilistic lethargy that he perceived in modern scientific culture. Moreover, they show that Nietzsche sees life as a drive that generates various ideological and normative structures—here Christianity, Schopenhauer's philosophy, Socratic and Platonic ethics. Moreover, the last sentence here, that this affirmation is "most rigorously verified by truth and science," offers a key clue into Nietzsche's appropriation of Darwin and the life sciences.

Nietzsche, Darwin, and the Meaning of Life

The passages throughout Nietzsche's writings that have been collected thus far have shown various contexts out of which the meaning of Nietzsche's concept of life can be reconstructed and have built a cumulative account of its place in Nietzsche's thought. The concept of life allows Nietzsche to unify claims about human psychology and affective life, value, art, religion, and science as well. Life is described as a physiological condition of the instincts, as the strength of the will that refuses the need to justify itself, as an affect of "joy" (*Lust*), and, most importantly as a drive that is naturally self-desiring. Moreover, it is used as a meta-philosophical principle. The drive of life is marshaled to show how Western philosophy began and why philosophical worldviews come to attract widespread assent. It explains why metaphysical conceptions have emerged that probe the ultimate nature of reality and yield ethical values in line with this reality, and why they have captured human thought and agency as matters of truly *vital* concern.

However, next to these philosophical meanings of the term *life* stands the obvious biological meaning as a designation of the general domain of nature that includes what we recognize as life-forms—plants, animals, insects, bacteria—of which humans are a part. Nietzsche's Life-philosophy cleverly plays with both the biological and philosophical meanings of the term in a way that reveals core problems that have occupied both present interpreters and his contemporaries. Life delimits the domain of nature studied by the biological sciences as a whole; it has to do with those self-organizing, self-maintaining entities, organisms, that have the capacity to grow, respond to stimuli, adapt, metabolize energy, repair themselves, and, finally, reproduce—all qualities that Nietzsche, often metaphorically, references when writing generally of life. Starting with the two-volume book *Human, All Too Human* around 1880, after his departure from the academic world and his position in Basel, Nietzsche's writings more and more came to invoke biological concepts that were emerging from the life sciences, but they did so in a way that made them speak to his earlier philosophical interest in the question of the value of life.

This is clear in his critical, but clearly appreciative, reception of Darwin. Nietzsche's famous passage on Darwin in *Twilight of the Idols* uses the term *life* to refer to a general feature of biological reality, showing at once his appropriation of the evolutionary conception of human origins and his unique challenge to what he saw as a basic tenet of Darwinian theory. The passage, titled simply titled "Anti-Darwin," reads, "The general aspect of life is not need, hunger, but much more abundance, opulence, even the absurd wastefulness—where one fights, one fights over power."[25] Andreas Sommer has used the compiled bibliography of Nietzsche's personal library to argue that in all likelihood Nietzsche had not actually read Darwin himself.[26] But, Sommer shows that he nonetheless possessed a general understanding of Darwinian theory from German thinkers that he did read carefully, as his marginal notes attest.[27] These included the philosopher of the unconscious Eduard von Hartmann (1842–1906), the early neo-Kantian critic of materialism Friedrich Albert Lange (1828–1875), and especially the German entomologist William Henry Rolph (1847–1883). Indeed, Heinrich Rickert also mentions Rolph's influence on Nietzsche's Darwinism in his 1920 book on *Lebensphilosophie*.[28] Nietzsche's marginalia and marks in a volume of Rolph's *Biologische Probleme zugleich als Versuch zur Entwicklung einer rationallen Ethik* (Biological problems together as an attempt to develop a rational ethic) in his personal library indicate that this "Anti-Darwin" passage drew on Rolph's challenge to what he saw as Darwin's view of the basic "drive" of living things.[29] Rolph argues that the "struggle for life" taking place in the natural world is not *for the sake* of survival or self-preservation—it is not a defensive struggle whose final cause is self-preservation—but rather an offensive struggle that is aimed at growth, expansion, and more incorporation by the organism of its environment into itself. Moreover, Rolph's work is an explicit attempt to elaborate an account of ethics that drew from knowledge of the biological world by capturing a biological principle of agency. Sommer shows that Nietzsche's use of Rolph displays his own "predatory" handling of sources by simply taking Rolph's aggressive drive to be an expression and validation of his conception of life, which he would later identify with the "will to power."[30]

Nietzsche's agreement with Rolph's critique of Darwin was itself an expression of the validation that he found in what he saw as Darwin's *agonistic* understanding of the evolutionary process.[31] But both this critique and Rolph's were made against the background of a general acceptance of Darwin's transformationist understanding of the origin of species—their descent with modification through common ancestry and conserved core traits. And this was shared by other German biologists that Nietzsche read carefully. Gregory Moore has catalogued Nietzsche's reading of Rolph and two other "anti-Darwinian" evolutionary theorists in Germany: embryologist Wilhelm Roux and cytologist Carl von Nägeli.[32] Wilhelm Roux argued against Darwinism, saying that the functional complexity of organisms cannot be explained by the *external* environmental pressures of natural selection alone and instead must involve also an *internal* struggle among parts of the organism for functional control over the whole organism that occurred during ontogenetic development. Carl Nägeli argued for a teleological "perfection principle" that drove the evolutionary process and so led to increasing progress through natural history.[33] None of these biologists objected to Darwin's central ideas of descent with modification that led to genealogical ancestry among species, or of gradual and adaptive evolutionary change by natural selection. Indeed, their theories are only intelligible against their acceptance of this broadly Darwinian background. However, what they did object to were the specific kinds of *teleology*—notions of what biological entities were striving after—that were encoded within accounts of this process. These biologists, and Nietzsche too, sought to identify different teleological principles that were driving biological organization and activity. Nietzsche intervened in these debates over the ends inherent in life-processes with his conception of the "deep, dark thirst" that he had characterized long ago in the second *Untimely Meditation*.

Before his "Anti-Darwin" passage in *The Antichrist*, Nietzsche voiced the same challenge to Darwinian theory in his 1886 *Beyond Good and Evil*, writing that "physiologists should rethink their assessment of the drive to self-preservation as the cardinal drive of an organic being. Something living wants foremost to discharge its power—life itself is will to power: self-preservation is only the indirect and most common consequence of this."[34] When we read this reference to the biological meaning of the term *life* as the basic drive of living things together with the passage about "tragic philosophy" quoted earlier from *Ecce Homo*, in which the Dionysian affirmation of life is characterized as "fullness," and indeed "overfullness," the systematic connection between the biological and ethical themes of Nietzsche's work that I am arguing for becomes clear. The concept that allows for the connection between the biological world, ethics, aesthetics, and religion is the concept of a teleological drive that defines living activity. This drive is aimed at its own satisfaction and at obtaining the external conditions of its satisfaction. Moreover, the drive that Nietzsche comes to see as the cardinal drive of the biological world is reflected in the aesthetic affect brought about by tragic drama—it is "the will to life rejoicing even over the sacrifice of its highest types." Tragic drama reenacts the unrestricted, unsuppressed expression of this inexhaustible natural *eros* of life for itself. In his objection to Darwin's alleged misunderstanding of the basic drives of organisms, Nietzsche recovers and reinserts into the evolutionary "struggle for life" the affect that both created and was expressed in the tragic *Weltanschauung*. His notion of the true *telos* of living things, as he understood it, was thus also a recovery of life's own unreflective, basic, and vital valuing of itself.

Nietzsche conceived life as *autotelic*—that is, directed at itself as the source of its own satisfaction. It was this autotelic nature of life that allowed Nietzsche to make the key move from description of a natural drive to discussion of the sources and criteria of ethical value and, further, to the project of a "revaluation of value" that characterized his final writings. Life desires itself, and only life itself is able to satisfy this desire. So the affirmation of life captures what constitutes the genuine fulfillment, satisfaction, and flourishing of a biological entity. Nietzsche's appropriation of Darwinism transformed his recovery of tragedy into a project of recovering nature's own basic affirmation of itself in a contemporary culture in which this affirmation appeared, to him at least, to be absent. His project was thus inherently evaluative at the same time that it was a description of a principle that explained the nature and behavior of organic forms.[35] The teleological character of life also yielded criteria of its flourishing. Indeed, the study of life fused metaphysics (general features of reality) and ethics (the ends humans set for themselves) in a way that allowed it to offer a constructive answer to Ogden's "existential" question posed in the introduction concerning the "meaning of reality *for us*." The biological world had an ethical and existential meaning for human life—in Nietzsche's view—because in it the nature and source of human satisfaction could be discovered. The key question, then, for the prevailing metaphysical and ethical systems became whether the ends that humans set for themselves were in harmony or in conflict with the ends of life.[36]

Life and Value

Nietzsche explicitly connects his biological notion of life to his reflection on values and valuing through his reflection on Socrates, a figure who also bridges his early and later writings. In *Twilight of the Idols* (1888), Nietzsche criticizes Socrates for three main reasons: his overestimation of reason and of the method of philosophical dialectic, his equation of happiness and virtue with rationality, and his status as a "savior" in Greek culture, offering a kind of existential cure for a condition of frustrated desire.[37] He goes on to repeat the same meditation on Socrates that first appeared in his book *The Gay Science* (1887); the scene of Socrates's death and his deathbed pledge that he owed Asclepius, the Greek god of medicine and healing, the sacrifice of a rooster were proof that Socrates was actually tired of life, that he considered the end of life a cure to be thankful for.[38] In the *Twilight of the Idols*, Nietzsche augments this judgment with meta-ethical reflection on what kind of truth-status such normative judgments on the value of life can have. He writes, "Judgments, value-judgments on life, for or against, can in the end never be true: they only have value as symptoms, they come into consideration only as symptoms—in themselves such judgments are foolishness. Accordingly, one must thoroughly stretch one's finger and make the attempt to comprehend this stunning finesse, that the value of life cannot be appraised."[39]

This passage announces the tension that hovers over the interpretation of Nietzsche's thought. On the one hand, Nietzsche here appears to deny that judgments about the value of life have any truth or falsity. He evaluates these judgments "for or against life" as "symptoms" of an underlying affective or psychological condition of desire.[40] On the other hand, Nietzsche clearly wants us to affirm, and even "crave" life. Socrates's equation of happiness, virtue, and rationality was not *false*; rather, it arose from world-weariness, existential fatigue, and resignation, which Nietzsche observed in Socrates's last words. This view of these judgments as symptoms suggests that to consider whether they are true or false is to apply the wrong categories. Instead, Nietzsche turned attention to the underlying condition that gave birth to these value judgments and focused on this condition as the rightful object of evaluation and critique. Socrates's judgment of life was not false, but it was defective in another way and according to a different criterion: it was unhealthy.

Later in the *Twilight of the Idols*, as Nietzsche turns his attention to the topic of morality as "anti-nature," the problem of the value of life resurfaces, this time detached from the figure of Socrates. Nietzsche writes, "When we speak of values, we speak under the inspiration, under the optic of life: life itself forces us to establish values, when we establish values, life itself values through us; . . . from this it follows, that also the anti-nature of morals, which comprehends God as the antonym and condemnation of life, is only a value-judgment of life—which life? Which kind of life?—but I have already given the answer: the declining, the weakened, the tired, the condemned life."[41]

In this pregnant passage, Nietzsche identifies life itself, that "deep, dark desire," as the source of human value judgments. Indeed, life again resurfaces here as a meta-philosophical principle. Ethical and metaphysical ideas, including conceptions of the divine, are to be understood as secondary effects of the underlying drives of those in whom they originate and whom they persuade and capture. The drive of life is not only the source of values, but also the source of the *necessity* of valuing: since it is a compulsion, it forces one to value. But, through all of the conceptual cloaks, cultural and biological forms in which life manifests itself, it is only the condition of this underlying, animating drive that cries out for recognition and gives us a way to assess, finally, what value these forms possess.

The two passages above illustrate the core of Nietzsche's biological approach to values. Such values are to be viewed in the way that a doctor treats one's vital signs—as indicators of "health." The question of whether or not life *really is* valuable independently of the acts of valuing that living things engage in—that is, whether they are *objective*—simply does not appear as a relevant question for Nietzsche. But the irrelevance of objectivity in this sense of independence is motivated by the fact that values only have meaning in relation to the living drives that they express and in which they manifest themselves. Values are not intelligible apart from living activity; they are not, in the end, about the world but about life itself. The normative criteria Nietzsche uses to assess value judgments are no longer *truth* and *falsity*, nor rightness and wrongness; rather, his criterion is the vitality of the basic desire of life for itself.[42]

In this claim we again find the expression of and justification for Nietzsche's rejection of philosophical rationalism and traditional philosophical modes of argument and reasoning, typified in Socratic dialectic. The idea that one can understand what values are and normatively evaluate them by appeal to reason alone was a philosophical mistake. The game of giving and asking for reasons was just that, a game, and it was impotent in relation to life. Nietzsche's critical method was to target life directly, to incite those underlying drives, affects, and desires that supported entire ideological edifices and, through this, to rediscover and recover life's own self-affirmation.

A Natural History of Christianity

Up to this point, religion has only been mentioned in passing. This is because the theoretical background mentioned thus far is necessary for understanding what Nietzsche takes religion to be, how he is able to subject it to critical evaluation, and how he relates it to science. While almost absent from his early work, Christianity becomes a persistent target of criticism in Nietzsche's middle and later writings, beginning in the 1880s. Indeed, as we have seen in the preface later appended to *The Birth of Tragedy*, Nietzsche retrospectively wrote Christianity into this early work as one of its main targets of criticism, even though it is mostly absent from the original text. The new preface describes Christianity as "the most excessive execution [*Durchfigurirung*] of the moral theme," which "allowed only moral values to be valid" and against which Nietzsche advances the famous idea of a "purely aesthetic justification" of life.[43] In his retrospective preface, Nietzsche links his later preoccupation with Christianity and moral values to his earlier efforts to recover the tragic, Dionysian affirmation of life from the Socratic dialectician and "theoretical man." As in these prefaces, we must sift through Nietzsche's polemical engagement with Christianity and its central values in order understand his theory of what religion *is*.

Nietzsche's critical engagement with Christianity is most systematic and vitriolic in two late works: *On the Genealogy of Morality*, published in 1887, and *The Antichrist*, written in 1888 and published in 1895 after Nietzsche's breakdown. Despite the individual aims and subtleties of each work, both contain variations on the same core thesis—the values embedded within the Christian tradition, ones that Nietzsche labels in general "morality," fundamentally reject the finality of life's value. The moral criteria of valuation express an underlying world-weariness and pessimism, a rejection of nature, and a weakened form of life's basic self-affirmation that Nietzsche himself wanted to revitalize through a recovery of the "everlasting joy [*Lust*] of becoming."[44] I concentrate in the remaining sections on Nietzsche's *On the Genealogy of Morality*, which is one of his most influential works for contemporary philosophers because it outlines a naturalistic, Darwinian, theory of religion in rich detail.[45] This work is especially important for this study because it is here that Nietzsche explicitly and systematically connects his reflections on life and value to the relationship between religion and science.

Nietzsche's *Genealogy* belongs in a tradition of modern naturalistic theories of religion that began with David Hume's *Natural History of Religion* in 1757. Like Hume, Nietzsche sets himself the two central tasks that such a naturalistic account must accomplish: first, it has to account for the *origins* of religions, and second it has to provide an explanation for their *persistence*. Nietzsche's *Genealogy* does just that, but it differs from Hume's treatment in decisive respects that relate to the preceding discussion of what religion is for Nietzsche. Hume treats religions as sets of beliefs about supernatural agents that arise out of fear and anxiety in the face of overwhelmingly powerful and opaque natural forces that humans are ignorant of and wish to control. In this conception, then, religions are anthropomorphic, pseudoscientific explanations of natural events that are born out of ignorance of the *vera causa* and so are superseded once better and more rigorous causal explanations are possible. Nietzsche rejects this as an overly scientific view of religion and instead treats religions as evaluative attitudes toward the goodness and value of life. Religions, he thinks, are not simply responses to fear or attempts to explain cause and effect relations; rather, they are sets of values that are generated by the same creative, "deep, dark desire" that drives living activity throughout the biological world.

This difference between Hume and Nietzsche is also reflected in the way each thinker approaches morality. Hume treated morality as a product of an inborn sympathy and natural social instinct of humans for fellowship with others. This sentimentalist moral theory also resonated with Darwin's later account of the moral sense as a natural, inborn other-regard that could have been favored by natural selection because more cooperative groups had a competitive advantage over other groups in the struggle for survival, resources, and reproduction.[46] Against this supposedly "superficial" explanation of the psychological underpinnings of moral values offered by "English psychologists," Nietzsche saw himself as penetrating deeper and farther into the natural sources of values by showing that moral values are expressions of the creativity of life as it seeks enhancement and flourishing.[47] This same creative drive was a principle of unity among the diversity of living things as they all undertook the "struggle for life."

Nietzsche's *Genealogy* is an attempt to solve a central puzzle raised by this conception of religion and his Life-philosophy in general: If all living things share a fundamental drive that is naturally self-desiring and self-affirming, and this drive is also the source of valuing, how did it become possible for a subset of living things, human beings, to reject or deny this drive, to reject the ultimacy of the value of life?[48] How can values arise out of such an allegedly basic and universal drive that are antinatural and no longer serve this biological *eros*? As the general argument of Nietzsche's famous Darwinian tale of the origin of life-denying values runs, the fall from the so-called state of nature, in which humans were naturally life-affirming, occurred through a successful revolt of the weak and socially powerless over the strong and dominant. This revolt was carried out by those who were subordinate in social rank and were full of *ressentiment* against their superiors, envying these superiors and yet simultaneously blaming them for their condition of frustration. Nietzsche postulates that in the evolutionary history of the human race prior to the birth and spread of morality, the noble, dominant individuals flourished amidst the harsh realities and conditions of life, while the subordinate individuals suffered.

The revolutionary birth of life-denying values contained in religious-moral systems occurred at the level of instincts, but it came to manifest itself in philosophy, in theology, and in the foundational values of social life generally. The birth of morality was the event in the natural history of the human species in which the will turned itself against life, and those individuals who rejected their conditions banded together to denounce and overthrow their life-affirming superiors. New, distinctively *moral* concepts of "good," "bad," and even "evil" emerged that no longer expressed the basic desire for itself that characterized natural life.[49] In this moment of the birth of religious-moral values, the Socratic *Geist* (mind) of "theoretical man" with its own *autonomous* values emerged from *Natur* (nature). But for Nietzsche this was not a triumphal emergence. It was precisely in this separation that antinatural values came to dominate, which, because of their very independence from and rejection of life, could find no satisfaction in the living world out of which they emerged.[50]

How could the moral revolution have been successful? While it was clearly political and social in scope, the revolt that Nietzsche envisioned against the natural, life-affirming individuals was first an ideological or philosophical revolution. This revolt was thus a demonstration to some extent of the power of the ideal over the real, as it attempted, by challenging and creating new sets of ideals, to change nature and bring it into conformity with reason. It was carried out by placing into question and revaluing the dominant normative order, and by spreading these new values and beliefs about the sources of value through society. The revolution was to overturn the previous values by claiming that indeed those who dominated and flourished under the regime of nature were "bad," while those who were frustrated and powerless, were "good." Explaining the success of this revolt was a matter of understanding how these values came to captivate, take hold, spread, and finally to appear normative, written into the very fabric and purpose of reality—even into the fabric of nature that they so clearly, according to Nietzsche, contradicted. Nietzsche understood the birth of Judaism, the birth of Christianity, and the birth of morality as original "revaluations of values" that created new values against which the natural drive of life seeking its own flourishing and enhancement was seen to be condemnable. Indeed, the highest aim became no longer the affirmation of nature, but rather the battle against life for the sake of something *other* and *better*. In this original revaluation, the so-called weak, weary, and frustrated wills became creative by reinterpreting the fact of suffering as a motivation to change the world, to produce new social, ideological, and material conditions in which they might find satisfaction.

In the first of the three essays comprising Nietzsche's *Genealogy*, the weak and the strong are understood as categories of rank in a primitive pecking order. But it is important to note that even here Nietzsche uses qualities of character and features of affective life, not those of mere physical strength and dominance, to differentiate the so-called weak from the strong. Nietzsche describes the nobles and the slaves in the first essay of the *Genealogy* through a list of what could properly be called virtues in ethical theory. The nobles are "thankful" (*dankbar*), "rejoicing" (*frohlockend*), "happy" (*glücklich*), "sincere" (*aufrichtig*), "naive," and "carefree," while the slaves are described as full of "*ressentiment*," "clever" (*klug*), "hateful," and "impotent" (*ohnmächtig*). Recalling his treatment of Socrates, here too Nietzsche shows that he wants to pinpoint the moment at which a certain condition of the instincts became creative and generative of new values. As dialectic came to appear to offer an existential therapy to those taken over by the allure of Socrates, here too Nietzsche argues that the new values of equality, pity, and the elimination of suffering took hold because they offered a form of therapy for a condition of frustrated desire and discontent. The irony of Nietzsche's account was of course that the so-called weak values eventually came to dominate the strong ones. But this shows us that Nietzsche's evaluative categories diverge in a crucial way from those of evolution and biology. *Strength* is not synonymous with Darwinian "fitness," and so it does not describe the values that spread in society at any particular time, nor does it describe those that spread the fastest and lasted the longest. Nietzsche's concept of strength and health attempted to capture conditions in which life flourished, and this is how it was at once a biological and ethical category. This diverged widely from the Darwinian criterion of fitness that is meant to capture reproductive success.

In the third section of the *Genealogy*, Nietzsche turned away from the Darwinian confrontation between weak and strong that spread moral values in a *quasi* cultural-evolutionary struggle for life to a psychological diagnosis of the instincts of those so-called weak types who first generated life-denying values. Nietzsche takes his most important discovery in the third essay to be the origins of what he calls the ascetic ideal. It is important to recognize the variety of activities that Nietzsche sees as manifestations of ascetic ideals, from science that aims at value-freedom to philosophy itself, morality, and even certain forms of art. But Nietzsche retains the core reference of the term *ascetic* to Hellenic and Christian religious practices of renunciation, self-discipline, and sacrifice of material well-being. *Askesis* means "training" or "exercise," and while originally applied to athletics in the Greek context, it took on a definitively religious meaning. Ascetic monks were those who rejected conventional markers of social success and material well-being through restricted diet or fasting, restricted sexuality, withdrawal from society, simple dress, and humble habitation. Nietzsche's repeated remarks about the "desert" and "escape to the desert" when speaking about ascetic ideals in the third section of the *Genealogy* are references to the desert father St. Anthony, who is one of the founding figures in the development of Christian monasticism and early Christian asceticism.[51] Anthony and early Christian communities that followed him into the desert epitomized ascetic ideals through rejection of what was perceived as the corrupting temptations of society; Anthony advocated a radical flight to the desert in the pursuit of higher forms of virtue, holiness, and flourishing. This kind of religious practice of sacrifice and renunciation, of controlling desire and cultivating oneself in pursuit of a higher form of human flourishing, is the paradigmatic meaning of the ascetic ideal. It is this kind of sacrifice and disciplining of life that Nietzsche finds at the origins of all of the so-called antinatural values of Christianity and morality.

The linchpin of Nietzsche's natural-historical genealogy lies in his analysis of the ascetic psychological type, for it is within the moral psychology of these individuals that natural life comes to turn against itself—and it spreads from there. Nietzsche's account of the ascetic priest is an attempt to resolve this paradox of how an antinatural rejection of life and a desire to transcend it nonetheless arise out of and persist in life. The key point is this: While the ascetic *appears* to seek an aim independent of the *autotelic* drive of life that characterizes all organisms, this project actually "originates from the protective and healing instincts of a degenerating life" as a "clever trick in the *preservation* of life."[52] This is the crucial turn in Nietzsche's natural history of religion, and it of course repeats Nietzsche's criticism of Socrates, whose death-bed confession showed, allegedly, that his philosophy came from a fundamental inability to affirm life. Nietzsche's diagnosis of the ascetic aims to show that values that reject the animating impulse of nature are not an *excep-*

tion to the *telos* inherent in life as such, but are a distinct expression of it. The crux of the *Genealogy* is the claim that human ideals and practices that appear to transcend and even reject the value of life for the sake of a different, higher form of satisfaction and fulfillment are indeed strategies for satisfying life's own "craving" for itself.

The ascetic priest, like the Socratic dialectician earlier, plays the role of a spiritual therapist who offers a cure for the sick-souled, who assuages an instinctual condition of what Nietzsche at one point calls the depressive listlessness (*Depressions-Unlust*) that expresses itself in ascetic values.[53] This cure, as Nietzsche explains, treats frustrated desire by attempting to block and suppress desire altogether. This results in a self-contradictory life, an internally divided appetite, for here is an "unsatisfied [*ungesättigten*] instinct and desire for power [*Machtwillens*], that wants to be master, not over something in life, but instead over life itself, over its deepest, strongest, and most basic conditions; here an attempt is made to use one's power to clog the sources of power."[54] The ascetic ideal is possible because the drive of life has turned against itself. And yet, the ascetic embodies the most ambitious and paradoxical form of this desire, one that takes on the impossible task of suppressing itself and, therefore, the condition of its own possibility.

Nietzsche calls the ascetic priest the embodiment of the projection of one's final satisfaction in a different world, somewhere-else (*Anders-wo*). His answer to the hermeneutic question that drives the third essay of the *Genealogy*, the question of what ascetic ideals *mean*, comes in the form of situating them within natural history and within life. The meaning of the ascetic ideal is that it gives a task and an aim that motivates action in the world to produce conditions that would allow a frustrated, embattled, and internally divided desire to be satisfied. The *telos* of this drive thus becomes transformed into the construction of another, not yet actual, state of the world in which life might be affirmed. Nietzsche writes, "The ascetic priest is the desire made flesh to be different, to be somewhere else, and indeed the highest degree of this desire, its actual zeal and passion: but of course the power of its desire is the chain that binds him here; indeed, he thereby becomes the tool that must work toward producing better conditions for being here and being human."[55]

By turning the will upon itself in this way, treating a frustrated desire by attempting to block desire altogether, the ascetic preserves life, indeed she must constantly and actively engage this most fundamental drive in a self-perpetuating, but ultimately self-frustrating, cycle. Nietzsche interprets ascetic ideals as treatments of the symptoms and not the underlying condition. The dissatisfaction that gives rise to ascetic values is also one in which the condition of its own possible satisfaction and genuine "cure" is absent:

namely, the recovery of life's own most basic thirst for itself. This condition is what Nietzsche late in his life claimed that his authorship was designed to recover. His project of "revaluation" was aimed at reversing this original revolution so that the natural desire of life for itself, which he viewed as a basic condition of human fulfillment and flourishing, might once again be satisfied.

Asceticism in Religion and Science

One of the most subtle and incisive claims in the third essay of the *Genealogy* is its assertion that the drives originally operative in ascetic religious practices of sacrifice and self-renunciation are the same guiding, normative moral values present in the aims of philosophy and in the value-free scientific attitude. Nietzsche offers one of his characteristically provocative psychological diagnoses of the common motivations at play in these seemingly diverse and conflicting human practices:

> Does the entirety of modern historical scholarship display an attitude more sure of life and its ideals? Its most lofty claim now is to be a mirror; it rejects all teleology; it does not want to "prove" anymore; it disdains playing the judge, and therein possesses its good taste— it affirms as little as it rejects, it establishes, it "describes." ... All of this is to a high degree ascetic; but it is at the same time to a high degree nihilistic; one should not deceive oneself about that![56]

Here, the conflict between science and life that Nietzsche examined in the second *Untimely Meditation*, and in his early reflection on tragedy and Socratic dialectic, returns. Nietzsche argues here that the value-free scientific attitude is actually a more refined form of the ascetic ideal that rejects life, because life, as we recall, is the desire that "forces" and "compels" one to posit values. Rather than being "value-free," the concept of a neutral and *therefore* objective science is itself driven by an ideal—Nietzsche even calls it a "faith" (*Glaube*)—and a vital interest that justifies it, gives it meaning, and provides it with an aim to strive for.

Nietzsche's shocking thesis about science and religion is that, far from being the enemy of the Christian, moral-religious vision of life, science is actually driven and sustained by the latter's core ascetic project. Nietzsche's Darwinian Life-philosophy thus places science, and the meaning of science *for* life, within a natural history by showing that it presupposes life and the values, norms, and ideals that arise from it. As Nietzsche writes of science, "No! Do not come to me with science, when I am looking for the natural antagonist of the ascetic ideal, when I ask, 'Where is the antagonistic will, in which an antagonistic ideal is expressed?' For this science does not even stand nearly enough on its own two feet, from every way of looking at it, it needs first a value-ideal, a value-creating power, in whose service it is allowed to believe in itself—on its own it is never value-creating."[57] Far from science being the enemy of religion, science actually depended on the same value-creating power—life—that was the natural source of religions themselves. Just as the religious ascetic flight away from worldly well-being to forms of radical sacrifice and renunciation was driven by life turning against itself, so too did the cardinal value of science—the ideal of knowledge *for its own sake*—arise out of a turning away from life and its own self-affirmation to something *other* and *different*.

For Nietzsche, the question of the value of life—that is, of whether or not the natural order of the living world was good in a final and ultimate sense—was the central question asked by religions and to which religions offered answers. This is why the question of the affirmation of life is his tool of explanation and critical diagnosis of religions and of the relationship between religious values and scientific values. Nietzsche turned to the biological sciences to ask about the moral meaning of natural history, and so to raise the central question identified by Thomas Nagel's "religious temperament" and Schubert Ogden's "existential question" that we met with in the introduction. That is, what is the *meaning* of nature *for life*. His answer was that life contained within itself its own affirmation, but that specifically human life had come to reject this and to pursue values that were autonomous with regard to life. For this very reason, however, they could find no ultimate realization in the natural world. In Nietzsche's thought, the desire for life acted as a basic, foundational, and natural form of valuing that was pervasive in the organic world generally and that in humans was the source of both scientific and religious values. Nietzsche left his readers with the question of whether or not life's basic self-affirmation could be recovered in modernity and modern scientific culture and, so, of whether or not scientific values could find their source in a new kind of religious temperament.

Conclusion

This chapter has reconstructed a recurring set of themes in Nietzsche's works that bind together his appropriation of Darwin, his concept of life, his mediation of religion and science, and his theory of value. Nietzsche's attempt to excavate life as a basic, instinctive drive and desire from underneath the manifestly value-laden spheres of culture and religion was a great philosophical departure from the classical tradition of German Idealism, which defended the distinctiveness of human rationality and the autonomy of rationality, normativity, and agency with regard to nature. I have argued here that Nietzsche was able to bridge the natural and the rational by viewing biological life as a *teleological* drive, that is, as possessing an inherent goal that established natural conditions of flourishing for living things with which humans could either align themselves or reject.[58] Nietzsche considered this Life-philosophy to be validated by the Darwinian picture of evolution driven—in Nietzsche's interpretation—by the attempt of organisms to flourish within the "struggle for life." He appropriated Darwin's view for his own aims by challenging what he saw as its limited conception of the fundamental goal of living things and, to an important degree, by blurring the line between the intellectual (or spiritual, as in the German word *geistig*) and the material. The material drives of survival and reproduction were already forms of the drive for satisfaction that would generate mind-infused cultural projects like science and religion.

In the *Antichrist*, the book that Nietzsche envisioned as his own "revaluation of values," the question of the value of life is brought to bear on the problem of nihilism. Nihilism, we recall, refers to Nietzsche's sense that attitudes in the modern world had come to reject life as a final value. Nietzsche writes, "When one places the weight of life not in life, but in a "beyond"—in nothingness—then one has taken away the weight of life completely. . . . Everything in the instincts that is healthy, that is life-enhancing, that is future-directed, only excites mistrust. To live in such a fashion that life has no more meaning—that, then, becomes the 'meaning' of life."[59] Nietzsche's plea for an affirmation of life came out of his sense that it was the only realistic conception of human flourishing in harmony with the universe that expressed itself in religious and philosophical systems. In this sense, Nietzsche's thought was a plea to affirm the real *as* the ideal. But it was also the attempt to generate a realistic idealism that constructed a conception of human flourishing grounded in what sciences were saying about the nature of the biological world. It was this world to which humans ultimately belonged, in which they were to flourish, and that they were to affirm.

However, it is clear that problems remain for this strategy: how can Nietzsche claim that we *ought* to affirm a life that is "beyond good and evil," or even that we *ought* to pursue the type of flourishing that Nietzsche wishes for us? As Nadeem Hussain points out, even if we grant that life is driven to affirm itself, and that this offers a genuine form of human flourishing, this still does not entail that *we* ought to affirm it or pursue this form of flourishing.[60] This is an objection to the normative authority of nature from the perspective of practical reason. The point is that life cannot serve as a normative criterion for what we should do, what is best to do, or what we have reason to do without appealing to grounds and values that have an authority independent of life. This is an objection that attends any attempt to develop an evolutionary or biological ethics, and it will come up again when we examine the work of neo-Kantians like Simmel and Rickert.

Viewing Nietzsche's project in light of this objection helps to indicate how radical it is. By rejecting the notion that reason is the source of values and viewing erotic, appetitive life as this source, Nietzsche rejected the notion that there can be any purely rational "oughts" for life. He did not aim to give us reasons to value life or argue that we ought to value it, but to show that valuing life was simply as vital as a heartbeat. Nature was not the source of *rational* authority or normativity; it was the source of this vitality, and it was not up to human rationality to legislate to nature what it takes for a living thing to flourish or find satisfaction. Because of this, the notion that rationally grounded values could indeed guide us to finding satisfaction in life was thrown into doubt.

The way Nietzsche resolves the relationship between the natural and the normative is a tension that persists in his thought, and it divides contemporary interpreters. Moreover, it raises another foundational problem for Nietzsche's philosophy that arises from the beguiling question of self-reference.[61] This is the problem of the extent to which Nietzsche saw his own conception of life to be merely a product of his own value judgments, grounded in the condition of his appetitive instincts, or as a characterization of life *as it really is*. I have claimed that Nietzsche saw his conception of life to be validated by Darwin and the life sciences. However, this is in clear tension with the meta-philosophical principle that metaphysical and ethical views, religions, and even science itself are products of value judgments that are symptoms of underlying instincts. This contradictory tension, like the tension between the natural and the normative, pervades Nietzsche's thought and divides postmodern readers of Nietzsche, who see him as rejecting truth and objectivity, and their philosophical opponents mentioned at the outset of this chapter. This self-referential problem also divided some of Nietzsche's earliest critics.

The following chapters show how thinkers influenced by Nietzsche responded to his project, as it has been set forth here, and struggled to understand the relationship between cultural activities like science and religion, which aimed at realizing core values, against the background of biological drives and processes of evolution in which they took place. For his critics, Nietzsche was more philosophically important and challenging than Darwin himself, because he more explicitly aimed to draw out the philosophical implications of biological concepts for problems in epistemology, metaphysics, ethics, and religion. But the unresolved tension remains between Nietzsche's dismissal of the idea that value judgments can be either true or false, normatively valid or invalid, and his own plea for the value of life by appealing to life directly. It was this tension—proper to what can not only be called Nietzsche's Darwinian Life-philosophy but also his Darwinian "Life-religion"—that Nietzsche's early critics seized on as they struggled with understanding the meaning of the natural sciences for philosophy. It was by resolving this tension in different ways that they came to offer alternative answers to the existential question of nature's moral meaning.

2

Franz Overbeck: The Life History of Asceticism

Franz Overbeck is known mostly—if at all—as one of Nietzsche's closest personal friends. Their friendship began at the University of Basel in 1870, where they both started their academic careers, Nietzsche in philology and Overbeck in theology, and lived in the same residence—the *"Baumannshöhle"* as they called it—one floor apart.[1] Outside of German scholarship, there have been very few studies of Overbeck's own fascinating struggle with the relationship between religion and science, which was informed by his critical appropriation of Nietzsche's conception of life.[2] However, while Overbeck was clearly influenced by his "brother in arms" at the University of Basel, he was no "mere" imitator. Overbeck was a careful and widely respected historian of the New Testament and Christian origins, and for this reason he was in a much better position to evaluate the scientific merit of Nietzsche's Darwinian genealogy of religion, his "just so story," against the historical record of the rise of Christianity in antique and late-antique culture.[3] Overbeck's analysis of the early development of Christianity was at the same time a study of the precarious place of historical scholarship within a faculty of theology.[4] Although he eventually became a trenchant critic of Nietzsche's appropriation of Darwin for sorting out the relationship between religion and science, Overbeck too was taken in by the lure of biological thinking for understanding cultural history, especially for describing the fate of religious worldviews in their own birth, development, maturity, and decline in history.

Overbeck's reflection on these topics emerged from a struggle to understand the contemporary culture wars over the place of Christianity in modern Europe that surrounded him and shaped his academic discipline. On the basis of his reflection on the development of early Christianity and its relation to Greco-Roman and Jewish culture during this early development, Overbeck came to conclude that there is a fundamental conflict between a so-called strictly academic and scientific perspective on religion and a perspective guided by so-called vital interests. He too developed a concept of life to capture the difference between these two kinds of interest, and he used this to explain the fate of religions in history. However, Overbeck's concept of life was developed both in critical response to Schopenhauer and Nietzsche and to teleological theories of cultural history that Overbeck confronted in the towering legacy of Ferdinand Christian Baur (1792–1860), the Hegelian historian whose work dominated early nineteenth-century scholarship in the history of religion. Baur and the Hegelians also saw dynamics of cultural change and the teleological development of history to be deeply entangled with structures of biological life generally.[5] But Overbeck's biological thought steered a unique path between both Schopenhaurian and Hegelian schools, and so offered a different and unique account of the relationship between science and life.

Overbeck was a voice from the fringes of mainstream, nineteenth-century German intellectual culture and theology in both geographical and ideological senses. He was an outsider to both liberal and conservative Protestant camps that dominated the politics of religion in Basel, which were divided precisely on the issue of how theology was to position itself with respect to modern science and secular philosophy. His professorship in the theological faculty of Basel afforded him proximity to the German academic tradition of history of religion in which he was trained, yet it also gave him a position from which he could safely voice an unconventional and deeply critical perspective on German academic, political, and popular culture without threatening his academic career. As Lionel Gossman notes in his book on the intellectual culture of Basel in Overbeck's time, the city's long history as an independent, sovereign city-state just on the Northeast border of modern Switzerland and pinched between France and Germany made it a perfect soil for the "unseasonable" antimodernism and cultural critique that some of its most famous inhabitants cultivated—of which Nietzsche is perhaps the most famous.[6] Cultural historian Jacob Burckhardt, historian of antiquity and founder of gender theory Johann Jakob Bachofen, Friedrich Nietzsche, and Franz Overbeck were all would-be reformers of their respective disciplines and critics of dominant elements of modern, liberal culture that had come to see a marriage between science, democracy, and liberal values as the key to human progress. However, while Overbeck was antimodern in his critique of liberal dreams of a harmonious marriage between science, cultural values, and a religious "ideal of life" (*Lebensideal*), he was nonetheless a staunch and loyal defender of the core liberal value that he thought was at basis of the modern scientific enterprise: freedom.[7]

The impact of Overbeck's thought, both in his time and in ours, was not only hindered by his fringe geographical and ideological place. Overbeck himself did not do much to increase the likelihood that his philosophical writings would change the course of the intellectual currents in his time. His published works consist mostly of strictly academic, historical studies of early Christianity, and his one major philosophical work from 1873, *Ueber die Christlichkeit unserer heutigen Theologie* (*How Christian Is Present-Day Theology?*), made little impact on the theological culture that it sternly criticized.[8] After this strident but stillborn attack of the foundations of theology as an academic and scientific enterprise, Overbeck confined most of his more philosophically targeted reflection to unpublished notes that he hoped would one day become a *Magnum Opus* that he called the *Kirchenlexicon* (Lexicon of the Church). The unpublished notes from the *Kirchenlexicon* that we possess offer a wealth of idiosyncratic critical reflection on topics ranging from Kant, the church fathers, and Pascal, to scholasticism, science, and methodology in history and theology. This work was to be a compendium of what Overbeck called "profane church history," and it was in these notes that his biological theory of religion and value is most explicitly stated.

The fact that Overbeck did not publish these notes is an indication of the peculiar mix of radicalism and reserve, one might even say quietism, that both his temperament and his writings possess. This precarious blend is obvious in the way Overbeck reconciled his vocation as a historian in a faculty of theology who remained in this discipline despite his sense that the perspective of a historian was deeply problematic for theology. This stance left many of Overbeck's critics dumbfounded. But this was no hypocrisy; Overbeck exemplified and defended the ascetic reserve of a certain form of value-neutral and positivistic science, of pursuit of facts independent of constructive ethical, cultural, or religious aims, and he saw academic *Wissenschaft* as the torchbearer of the Enlightenment ideal of freedom and autonomy. The boundaries between science, philosophy, and religion were fixed not by a positivist division between facts and values, but by what Overbeck took to be the strangeness of science when viewed from the perspective of organic life out of which it emerged. Despite Overbeck's commitment to the pursuit of knowledge *for its own sake* and his attempt to understand the cultural meaning of this unique project, the aggressive tone of his philosophical writings shows him leaping beyond the strictures of academic *Wissenschaft*, even into the art of polemic as it was so masterfully practiced by his friend Nietzsche. It is in these moments of transgression that Overbeck demonstrates the ways in which *Wissenschaft* too might be a product of vital interests.

While never achieving much status within the catalogue of great nineteenth-century thinkers, except through his friendship with Nietzsche, Overbeck has had a subterranean impact on twentieth-century thought. His work has been engaged by some of the most influential thinkers in twentieth-century philosophy and theology, including Ernst Troeltsch, Martin Heidegger, Karl Löwith, Walter Benjamin, and Karl Barth.[9] Heidegger read Overbeck intensively during his confrontation with theology in the early 1920s, and he was impressed with Overbeck's critique of what he saw to be an unfortunate accommodation of the early Christian ascetic impulse and critique of worldliness to modern culture in general.[10] While Troeltsch saw Overbeck's strict separation between science and life, history and religion, as a mistake, recent scholars have shown that some of Troeltsch's most influential writings were conceived in part as ripostes to Overbeck's strict division between history and theology.[11] Barth, on the other hand, could only concur with Overbeck's critique of the turn to history ushered in by nineteenth-century liberal theology. While Barth clearly ignored Overbeck's rejection of attempts to place theology on any foundation other than life, he found Overbeck's division between scientific history and religion to be a decisive move in favor of his rejection of history and turn to the immediacy of encounter with scripture. Overbeck's original articulation of the division between science and religion thus prefigured and informed debates that would shape twentieth-century theology and that still have resonance for theological debate today.[12]

The *Antrittsvorlesung*: Outlines of a Concept of History

As academic custom dictated, Overbeck was required to give an inaugural lecture upon his assumption of his first academic position in New Testament and Church History at the theological faculty in Basel in 1870. The topic that Overbeck chose for his lecture was an important political move for reasons that I explain below, yet it also raised philosophical issues at the heart of this book that would continue to occupy him for the rest of his career. The lecture was held at the beginning of Overbeck's tenure in Basel, before he had much interaction with Nietzsche, who would soon be his neighbor, and so is also evidence that these central themes of Overbeck's thought were not merely absorbed from his "brother in arms."[13] The title of Overbeck's inaugural lecture, "On the Emergence and Right to a Purely Historical Consideration of the New Testament Scriptures in Theology" (*Über Entstehung und Recht einer rein historischen Betrachtung der Neutestamentlichen Schriften in der Theologie*), directly announces the crucial problem of how to understand the relationship between genealogical and normative inquiry, between Kant's question of fact (*quid facti*) and question of right (*quid juris*) as it applies to the relationship between the distinct intellectual tasks of science and theology.

The genealogical story of Overbeck's *Antrittsvorlesung* is quite straightforward: it was not until the early nineteenth century, through the work of the famous Hegelian church historian at Tübingen, F. C. Baur, that a so called "genuinely" historical perspective on the earliest history of Christianity had been achieved. Moreover, Overbeck argues that the emergence of a scientific consciousness of history was itself a historically novel, momentous achievement of utmost consequence for theology. His lecture provided a genealogy of historical consciousness by tracing its emergence in some of the earliest and most influential Christian writers—Irenaus, Clemens of Alexandria, Tertullian, and Origen—and moving through the Middle Ages and finally to the Reformation. Overbeck contends that what is most clear in telling this history is the absence of what he calls a "purely scientific interest" in the New Testament narratives and the events and dynamics that shaped the beginnings of the Christian movement up to the formation of a Biblical canon.[14] This absence was conspicuous, since it was the "purely scientific" perspective that modern theology was struggling to come to terms with in his time. Overbeck's turn from genealogy to questions about what the scientific perspective was and what it meant for theology was a turn from history to contemporary cultural critique.

It might seem strange to contemporary ears that a Hegelian historian such as F. C. Baur was considered "purely scientific," let alone by a thinker such as Overbeck who, as we will see, rejected the teleological assumptions of Hegel's philosophy of history. But our judgment is anachronistic, and it is important to recall that Hegel's conception of dialectic was to capture natural, developmental dynamics of thought generally, and so also of the dynamics underwriting the development, change, and succession of cultural products of thought and agency. Of course, calling this developmental patterning and process of thought "natural" here does not mean that it is to be explained through physical or mechanistic causes. Instead, the term *natural* here refers to a dynamic process that unfolds according to its own internal principle, or "nature." The development of rationality and mind, or *Geist*, was such a process that mirrored natural processes of organic development. What was exceptional in the Hegelian perspective, for Overbeck, was that Baur's historical method, like Hegel's, did not posit supernatural causes external to the inherent principles of things—their "natures"—but sought to discover principles of cultural change that were internal to the unfolding processes themselves. The scientific method in the study of history was designed to capture an internal logic in the self-development of human self-understanding. For Overbeck,

F. C. Baur arrived at a "purely scientific" historical method because he self-consciously approached the history of religion in terms of the same inherent, immanent principle that governed the self-development of all mental activity, and indeed of all biological form in general. In the same way that Hegel assumed an inherent principle of self-organization within the development of a seed into a mature tree, so too Baur argued that the history of religions followed a teleological pattern of development from the early "seeds" of religious movements through their various cultural changes and into their mature fruition.[15]

There are many common misunderstandings of this Hegelian approach to history that especially revolve around the notion of teleology discussed in the last chapter. Hegel and Baur did believe history was teleological, but the teleological character of history was derived from a directed character that *Geist* was thought to share with all organic life. In the same way that the growth of a seed was thought to be "directed" toward its mature state as an oak, so historical change followed the same organic pattern of development. History was a process in which later stages of cultural development, which meant simply, in their terms, the development of mind (*Geist*), were in some sense necessary and more comprehensive outgrowths of previous stages. Yet the process in general was one by which humanity was to fully become what it is, or to realize its nature. In the organic, biological world, teleological development was understood through the life-history stages of birth, growth, and development into a mature individual; in the sphere of mind, it meant (for Hegel and Baur) coming to knowledge of the absolute—or what some religions called *god*.[16] Through working with the historical record, the Hegelian historian was to trace how *Geist* in fact historically and materially unfolded through a dialectical process of development. Baur's Hegelian teleology thus also involved a philosophical appropriation of biological conceptions of organic growth and development, but in challenging Baur, Overbeck challenged this way of applying biology to history, of transitioning from the organic to the rational, as well.

Understanding what Overbeck rejected in Baur and Hegel is key for understanding his later mediation between science and life. He rejected the foundational principle of a progressively greater self-realization of *Geist* that was the driving force in Hegel's and Baur's teleological conception of historical development. Overbeck well understood that this particular teleological conception had dramatic consequences for how these thinkers judged the relationship of the past to the present. The most dramatic consequence was the necessity of recognizing irreconcilable difference and *discontinuity* between past and present. The idea that the present is the culmination and fulfillment of the past, the necessary realization of what was incipient in past stages as what they were all-along striving for, had to be rejected. Instead, historical change required the recognition of radical break between the cultural forms of past and present. The present had to be understood to be something completely different from the past out of which it emerged, a new form of life or, in Darwinian terminology, a new "species." In Overbeck's case, this change in understanding the analogy between biological and cultural development had dramatic consequences for evaluating the relationship between modern culture and Christianity at its point of origin and for identifying just what a naturalistic, "purely scientific perspective" on the history of religions entailed for contemporary theology.

The characteristic marks of a "purely scientific perspective" that Overbeck identifies in this early essay demonstrate his break with Hegelian teleology. The presuppositions of the historical perspective on the earliest beginnings of Christianity, he writes, are few. First, this perspective involves the acknowledgement of *discontinuity* in the "difference between the beginnings of Christianity and the respective present."[17] Second, Overbeck writes that the historian works according to the general (one might say evolutionary) principle that "what is has become, and originally was not so."[18] He rejects the notion of eternal, unchanging substances or essences in both cultural history and biology, and while he does not cite Darwin here, this kind of historical consciousness about the lack of fixity in biological and cultural form was one of the most dramatic implications of evolutionary thinking. As Hegel had historicized philosophy, ethics, and religion, Darwin historicized nature. Third, and crucially, Overbeck refers to the historian's value-neutral interest in understanding the past. The historian is not interested in or motivated by judgments about whether what happened was "good" or "bad," but only what can be known about the precise chronological and causal relation of events as they occurred in time and as far as they can be known.[19] Thus, the scientific historian is defined by her paradoxical disinterested interest in knowledge *for its own sake*; the interest of the scientist begins and ends with knowledge of cause and effect, de-coupled from any judgments about the moral meaning of what we observe.

The crucial aspect of the "purely scientific perspective" on religion that emerges in Overbeck's lecture, which he took from Baur, is that religion is to be explained according to the same principles as all other cultural or natural phenomena, without positing any special transcendent causes operative in religions alone. Of course, this immediately placed interventionist conceptions of divine revelation, or notions of the presence of special divine agency operating through the development of religion out of bounds. Even though

Overbeck could acknowledge that the historian could neither support nor deny claims about supernatural agency, these claims were not necessary to account for the rise and spread of religious worldviews. The spread of religions, too, was to be understood as a human, all-to-human, phenomenon, and the development of religion in culture became one, albeit decisive, element of profane cultural history. This move of disallowing the presence of transcendence or the sacred in explanations of the history of religions, of course, placed a decisive wedge between history and theology.

Overbeck argued that the historical perspective could be further captured by the way it treats the Bible as a text. The historian views the Bible as a window into the animating life of the people out of which its central values and teachings emerged.[20] Cultural texts and material products were windows into the animating spirit of the peoples, and this view had clear parallels with Nietzsche's approach to Greek Tragedy discussed in the last chapter. Both of these were also indebted to the conception of cultural studies as a study of the animating spirit of cultural forms defended by Overbeck and Nietzsche's older colleague in Basel, Jakob Burckhardt. But, as we will see, Nietzsche and Overbeck gave this idea of animating life a distinctly biological meaning and grounded it the world of living things generally.

After Overbeck completed his genealogy of how a historical perspective arose within theology, he turned to the second, normative, question announced in the title of his inaugural lecture. This question of the *right* to a historical perspective within theology thus took shape as a question about whether the scientific interest in history could be understood to serve a religious aim. In response to this question, Overbeck argued that differences of *interest* and *aim* are ultimately decisive for distinguishing between the theological and the scientific perspectives. The scientific aim of the historian is to produce, on the basis of the existing evidence, greater and more accurate knowledge of events, attitudes, and practices of people who have lived and how they have affected subsequent events. But theology too is defined by its unique goal and conception of its own task. This goal, Overbeck claims, is to mediate between science and religion by constructively defending how these can be brought into rational harmony, or by policing the boundaries between the two. Theology is thus essentially *apologetic*, aimed at defense against secular critique and against external, foreign interests.

It should be clear from this analysis so far how close Overbeck's early lecture comes to the worries Nietzsche expressed in his second *Untimely Meditation* concerning the use of history *for life*. Indeed, Overbeck understands theology as a *task* whose central aim is to put science in the service of religious life, which is animated by core ideals and values. The key philosophical claim of this inaugural lecture emerges from Overbeck's characterization of the relationship between science and religion. The shift in interest that the purely scientific perspective brings, he claims, also brings a fundamental change in the *object* of investigation. Theology and science are simply talking about two very different things.

Bernard Williams's discussion of the distinction between history and myth in his recent book, *Truth and Truthfulness*, can help bring Overbeck's concern into focus for contemporary readers. While I introduce Williams here, it will be helpful to keep his statement of the problem in mind throughout the chapter, since it captures a point that animates Overbeck's later writings on science and religion. Williams reflects on the nature of a scientific perspective on history by turning to the ancient historians Herodotus and Thucydides. His aim is to understand and evaluate the claim defended by many scholars that Thucydides invented the modern, scientific concept of history. Williams aims thereby to ascertain what constitutes a distinctively scientific consciousness of history. He writes,

> Historical time provides a rigid and determinate structure for the past. Of any two real events in the past, it must be the case either that one of them happened before the other or that they happened at the same time. This does not hold for the mythical, or, more, generally, for the fictional or the imagined.... To say that a statement about an event is historically true is to imply that it is determinately located in the temporal structure; if it is not, historical time leaves it nowhere to go, except out of history altogether, into myth, or into mere error.[21]

For Williams, the scientific historical perspective arises over against what he calls the "mythical." This new perspective is constituted by an entirely different set of cognitive aims and questions that are not present for the mythical perspective; in this respect, the mythical perspective is marked by what one can call a certain naïveté. The decisive difference between the two perspectives lies in the kind of interest that motivates inquiry and the kind of questions it becomes appropriate to ask about stories and texts from and about the past. The point—for Williams and Overbeck—is that scientific history is a special kind of *task* constituted by cognitive interests that aim at the collection and chronological ordering of events in their causal relations. As a task, it is a normatively guided and constituted activity of constructing evidence and adjudicating claims, and Overbeck goes as far as to call it a "way of viewing" (*Betrachtung*). Moreover, Williams helps us see the key point of Overbeck's discussion: the concept of historical reality itself as a world-picture only becomes possible through the distinction between history and myth, and this required the emergence of a distinct set of cognitive interests and attendant norms.

The historical perspective establishes criteria for what counts as historical reality, and it thereby shows that it is constituted by normative elements. For example, one must be able to tell a causal story that links events to the time, place, and effects of other known events in a chronological series. For Williams, the decisive philosophical lesson of these reflections emerges when one asks whether or not ancient Hellenic cultures believed their myths to be true, or mythical figures to be real. This question as it is formulated from the perspective of scientific historians contains a telling misunderstanding that both Williams and Overbeck draw attention to. Because the normative scientific concept of history had not been formulated before such "purely historical" writers as Thucydides had discovered or invented it (Overbeck sees Eusebius as the first thinker to introduce historical consciousness into the Christian tradition, although he also regards Eusebius as the most unreliable and untrustworthy of Church historians), the question of the historical reality of myths *as we now ask it* was not an intelligible question.[22] The concept of historical truth was itself an historical achievement that first made it intelligible to ask whether or not the past recounted in stories was real. Williams writes further, "Such a practice [of telling stories about the mythical past] is not inherently unstable; it can last for long periods of time. But it becomes unstable if the kind of question that is appropriate to the everyday begins to encroach on the stories about the old days, and there ceases to be a natural and unreflective way of moving from one way of talking to the other."[23] Once this natural and unreflective way of relating myths to everyday events and causes and to the principles operative in the everyday becomes disrupted, it first becomes possible to distinguish between myth and history, and this raises the possibility of a conflict between the scientific world-picture and the normative underpinnings of religious life.

In the *Antrittsvorlesung*, Overbeck posed the challenge that historical *Wissenschaft* and its purely scientific interest in the explanation of the past calls into question its constructive use for the life of Christianity in a cultural context dominated by modern science. Science introduces a different interest, a different set of normative criteria for reality and inquiry into it, and so ultimately a wholly reconceived object of investigation. As a result, the question of the *justification* of the criteria of historical reality for theology no longer had an obvious answer. Overbeck's answer to this *quid juris* question gives a key clue into his general, genealogical, mode of analysis. He does not offer a transcendental defense of the authority and value of the scientific interest and its implicit criteria of knowledge and reality. Instead, he offers a genealogical argument: it was Protestantism that introduced into Christianity the conditions for a purely historical perspective to emerge and to become normative.

Overbeck's answer to the *quid juris* question of the right to the historical perspective in theology is, then, a version of a well-known "Protestantism and science" thesis that derives the value of science for theology from animating concerns of Protestant reformers. The right to the historical perspective within the program of theology, Overbeck argues, is a consequence of its emergence from the original Protestant commitment to the value of individual freedom and the authority of individual religious commitment. The historical perspective is an exercise in freedom because it allows one to distance oneself from one's given tradition or culture and place it in question, to no longer accept it as taken for granted, and to see it as contingent. This, for Overbeck, is an exercise in autonomy. The justification for this task within theology is a consequence of the genealogy of the principle of freedom out of the Reformation and the renewed emphasis on the Bible as the normative source of religious authority that was a driving force behind the Reformers themselves and their rejection of clerical authority over scriptural interpretation. The ideal of freedom that is at the root of the scientific investigation of value-neutral causal relations was thus also a product of the ideals of a religious tradition; in this case Overbeck claims it was an inheritance from the ideals of the Reformation.

There are problems with taking Overbeck's answer to the *quid juris* question of right in the context of his *Antrittsvorlesung* at face value. There were complex politics surrounding Overbeck's first lecture to his colleagues and those who had hired him in Basel. Overbeck was a compromise candidate between the liberal reformers and the orthodox conservatives, who themselves were divided over how theology was to respond to the advancing natural and historical sciences. We can thus surmise that he had a distinct strategic interest in mind in giving this lecture, even though this was tied to the philosophical commitments that shaped his understanding of history and theology.[24] Overbeck wanted to make it clear that he did not intend his work as a historian to serve the liberal reforming party's desire to peacefully harmonize religion with rationality and the developments of science, as they were expecting. Yet, he also needed to assure his colleagues that his work did have a continuing connection to the guiding aims of modern Protestant theology and culture. Overbeck's internal, genealogical justification of the value of history for religion draws on the already existing ideals and interests of his audience to justify the perspective of profane history that he aimed to put into practice.

Nevertheless, the message of Overbeck's inaugural lecture was quite radical. Overbeck charged theologians and the theological culture of his day with not taking the division between the aims of science and those of theology seriously enough, accepting artificial solutions that obscured how deeply the context of a scientific culture, with its own autonomous, normative commitments, challenged the aims of religious life. Although his genealogy of the historical perspective at times sounds like a triumphal narrative of the emergence of reason, freedom, and scientific criticism from his own Protestant cultural tradition, this would be a superficial reading. Overbeck was much more ambivalent about the meaning of this scientific perspective and what it signaled not only for the ideals of Protestantism but for the pursuit of ideals and values generally. The ominous final lesson of his inaugural genealogy of the historical perspective from the roots of Protestant theology ends simply with a warning not to underestimate the challenge that life is faced with in a world saturated with purely scientific interest.

Overbeck's *Streitschrift*: A Genealogy of Asceticism

While the tone of Overbeck's 1871 *Antrittsvorlesung* was (at least to some degree) moderate and accommodating—a continuing trace of the influence of Baur's Hegelian and mediating approach or simply a prudent political maneuver—his subsequent philosophical effort was a strident critique of the theological task of mediating between science and life. In Overbeck's 1873 *How Christian Is Our Present-Day Theology*, this mediating task is judged a *contradictio in adjecto* (contradiction in terms).[25] As the work's provocative title clearly states, Overbeck's aim this time was to reject not only the right to a purely scientific perspective within theology, but also the right to the mediating efforts of academic theology in contrast to what he saw as the most basic animating concerns of the Christian religion. Overbeck offers a telling and intriguing characterization of his goal as the attempt to provide a "theoretical solution to the problem of the relationship between Christianity and the educated mind [*Bildung*]."[26] There is much of interest in the assumptions that cause one to raise such a question, and Overbeck's own unique Life-philosophy emerges through his attempt to show the nature of this problem.

Overbeck's radical turn against the theological camps of his day rested again on establishing boundaries. This time the boundaries are not between science and theology, but between theology and religion. The *Christlichkeit* recalls the central themes of the *Antrittsvorlesung*: religion and science take radically different interests, questions, and aims as points of departure, and these shape their unique objects of investigation. Yet in this tract, Overbeck argues that the mediating task of theology also pursues an aim foreign to the impulses that animated the original formation of its founding religious values and ideals. And without these impulses, theology was meaningless. His historical research into Christian origins aimed to trace the genealogy of what he calls the *Lebensbetrachtung* ("view of life") of the earliest Christians.[27] And this view of life was something very different from theology. We have seen in the *Antrittsvorlesung* that Overbeck doubts whether or not *Wissenschaft* can serve the interests of religion, but his more controversial and radical view in the *Christlichkeit* is that the mediating efforts of theology are also questionable.

Overbeck, like Nietzsche, captured the two conflicting spheres of religion and science through the traditional categories of *Glauben* (faith) and *Wissen* (knowledge). Like Nietzsche, Overbeck does not directly define these terms, and so their meaning must be inferred from his criticisms of various theological rivals, to which he shifts his attention. Overbeck's writing of the *Christlichkeit* was prompted specifically by two works that he strongly objected to, which shared the common aim of revitalizing and transforming the Christian heritage of modern European culture to conform to a modern, scientific worldview. The first was a pamphlet written by the Orientalist, Bible scholar, German nationalist, and notorious anti-Semite, Paul de Lagarde (1827–1891). Lagarde called for a new, German national religion that could be produced out of a state-sponsored *Religionswissenschaft* (science of religion).[28] Lagarde's idea was that scholarship on the history of religion could be the foundation for a new religious worldview that could replace the stale forms of Christianity that no longer appeared to satisfy scientifically minded modern Germans. Scientific research into the history of religion could then become the basis for a new, national religion in which a unified German identity could take root. Overbeck's *Christlichkeit* was an attack on Lagarde's conception of a theologically constructive "science of religion," which for him amounted to nothing more than a religion invented by scholars—one that could in no way generate a set of values that humans could actually live by.

Overbeck saw the same stale and uninspiring attempt in a popular book by the infamous critical historian of the life of Jesus, David Friedrich Strauss. Strauss's 1872 book, *Der Alte und Der Neue Glaube* (The old faith and the new), was immensely popular and widely read. It defended a new, post-Christian spirituality and worldview (*Weltanschauung*), an updated surrogate for Christianity that would be more palatable to scientifically educated moderns and more compatible with modern science.[29] Strauss argued for a religious reverence for the universe itself as the scientifically appropriate expression of religious feelings in the modern age. Like Nietzsche, whose first *Untimely Meditation* branded Strauss's proposal as cultured philistinism, Overbeck found Strauss's "new faith" hollow and superficial. The period of

Overbeck's critique of Strauss was the period of Nietzsche and Overbeck's most intense intellectual exchange. Indeed, in 1873, Nietzsche presented Overbeck with a volume that included Overbeck's *Christlichkeit* and the first essay of Nietzsche's *Untimely Meditations*—entitled "David Strauss: The Confessor and the Author"—bound together with a unique inscription that referred to these two strident polemics as "twins."[30]

Overbeck complained in the *Christlichkeit* that both Lagarde and Strauss diluted the animating, vital concerns that originally generated the core *Lebensideal* (ideal of life) of Christianity. Their new, substitute religions yielded merely abstract *Gedankendinge* (thought-things) and not existentially satisfying or moving ideals at all.[31] In both cases, scientists were calling on themselves to imitate the inspired prophets of old and capture the *vital* interests of moderns trying to come to terms with the meaning of human life. But these would-be builders of a new religion failed to understand the vital concerns that sustained and generated the ideals of life that they wished to supersede. These concerns were decidedly "extra-scientific"—indifferent and potentially even hostile to the aim of truth *for its own sake*.

These criticisms of Lagarde and Strauss give initial clues into Overbeck's understanding of what it is about religion that leads to its conflict with both science and theology. In attacking the "mediating" projects of Lagarde, Strauss, the Hegelians, and other academic theologians in his day, such as Adolf von Harnack, Overbeck invokes the concept of "myth" to capture the religious perspective in a way that recalls Bernard Williams's discussion mentioned above. He writes, "Modern theology in particular is not at all in the position to produce anything that might even remotely resemble a religion. A religion can be completely indifferent to its store of myths as long as its myth-constructing power is still alive, that is, as long as the miraculous powers that brought forth its basic myth are still active in it."[32] Myths are "vehicles" for life, which is invoked here as a basic, animating power that generates them and that is expressed in them. In particular, Overbeck contrasts "living myths" (*lebendige Mythen*) with the constructions of theology and the sciences, which Overbeck soon goes on to say can function only as "surrogates" to prop up "myths that can no longer live on their own."[33] This use of the opposition between "living" and "dead," "active" and "inactive" forces out of which myths originate is central to Overbeck's distinctions between science, religion, and theology. Of course, it directly resonates with Nietzsche's concept of life as a value-creating power that was presented in the last chapter. Whereas scientific investigation is oriented by its aim of knowledge *for its own sake*, myths are the products of living "miraculous forces"; they erupt out of and serve a vital concern active at their point of origin and spread through an affective resonance. It is the continued presence of this vital concern that sustains them as effective forces in history, culture, and personal life.

This vital current that gives rise to myths is of a particular form. Later in the *Christlichkeit*, Overbeck characterizes the importance of myths for religion by drawing specifically on organic metaphors. He writes that "myths indivisibly belong to religion, which justifies these myths, and so it is not constituted by them completely; instead, myths do not allow themselves to be uncoupled from the specific view of the world and view of life (*Welt- und Lebensbetrachtung*) that, as stem to leaves, first produces them."[34] Here, Overbeck turns to biology, in particular to plant life, to explicate the way in which vital interests produce myths that embody and express them. Religions here are myths and stories that express a certain evaluative stance with regard to life and that are produced from this sense of value, just as a stem produces leaves. Yet, perhaps more crucially, this basic, underlying stance gives myths "living truth" (*lebendige Wahrheit*); the brute fact of this given vital concern is at the same time the criterion of their truth.[35] Overbeck's main criticism of both conservative and liberal theological camps in his day, which argued over how to mediate between science and religion, was that they had already surrendered the "kernel" of the vital forces that produced the Christian view of life in the first place and merely quarreled over the "husk"—its now empty myths, dogma, and creeds.[36] They had thus already surrendered a sense of value that was the source of religious normativity to newly formed, extra-religious interests.

Overbeck's vitalist conception of a view of life (*Lebensansicht* and *Lebensbetrachtung*), which he used interchangeably with the term "ideal of life" (*Lebensideal*), is central to his theory of religion.[37] But it is important to notice that Overbeck's vitalism is derived not only from reading thinkers like Schopenhauer and Nietzsche, but also from his own historical investigation into what the most original forms of Christianity were (*Urchristentum*). As a church historian, Overbeck's account of the Christian ideal of life stemmed from investigating early Christian writings in the context of the Jewish and Greco-Roman cultures in which they took shape. The notion that active "living forces" produce religions and sustain them justified the genealogical method of looking to their point of origin and birth to find their vital core. This historical investigation, in turn, confirmed that the animating pulse behind the birth and spread of the Christian religion was a distinct valuation of life that lay behind its myths and creeds. Overbeck's genealogical thesis— one expressed well before Nietzsche's genealogy—was that the original Christian view of life was a world-rejecting (*Weltvernichtend*) and world-escaping (*weltflüchtigend*) asceticism. He writes, "Surely there cannot be a more otherworldly faith than that of the earliest Christians in the imminent return of Christ and the demise of the world in its present form."[38] This dichotomy between worldliness and otherworldliness pervaded Overbeck's understanding of what the "genuine" Christian ideal of life was, as it went on later to shape Nietzsche's conception of life-denying and life-affirming instincts.

Nietzsche's later Darwinian *Genealogy* argued that ascetic ideals arose out of a world-weary fatigue and came to dominate through an ideological *coup d'état* of the powerless against the powerful. Overbeck's more historically credible counter-genealogy contends that the ascetic ideal arose from a transformation of an early apocalyptic expectation of the imminent end of the world that was prevalent in many Jewish sects into a more abstract withdrawal from secular culture and devaluation of its ultimate significance. Overbeck writes, "The expectation of the return of Christ, which was the offspring of altogether Jewish hopes [*judaistische Hoffnungen*] for the prospect of the decline of the world in its present form, was untenable in its original form and transformed itself into the thought of death, which already according to Irenaus should always accompany a Christian—into the *memento mori*. That Carthusain greeting summarizing the basic wisdom of Christianity more deeply than, for instance, the modern formulation that 'nothing disruptive' should 'be allowed to come between man and his primal source [*Urquell*].'"[39]

The essence of Christianity lay in its valuation of life, not in its metaphysical teachings or in its characterization of the human relationship to the "primal source" of being or to the observable universe, as Strauss had claimed in *The Old Faith and the New*. As Nietzsche later echoed, the basic impulse of Christianity was a form of asceticism or, more specifically, an expectation of the end of the world that expressed an ascetic "world-negating" valuation of life. For Overbeck, this original and literal apocalypticism of the Christian movement was later transformed into the ascetic rejection of worldly well-being for higher forms of flourishing. And so, the essential otherworldliness of the original view of life that drove early monks like St. Anthony to the desert would later ground the formation of ascetic monastic movements. The devaluation of worldly markers of success, well-being, and status was the basic impulse behind the construction of a new *Lebensideal*.

Overbeck's early *Streitschrift* argued that the ascetic ethic of sacrifice and renunciation in early Christian figures such as St. Anthony and the desert fathers was a "revaluation" of values that Christianity introduced into Jewish and Greco-Roman culture. This estimation of the origins of otherworldly asceticism is historically inaccurate, since it has precedents in Greco-Roman thought and culture.[40] But the issue of importance here is the philosophical background that leads Overbeck to look for the stance that a religion takes on the value of life to find its "essence," and to look at the origins of a religion to find the vital core of its teachings. While science seeks knowledge for its own sake, religions establish a set of teachings that express a fundamental attitude toward the value of life and a manner of conduct in thought and action—a "way of life"—that embodies and realizes this value.

Overbeck's conception of the basic conflict between science and a vital *Lebensideal* announced another crucial departure from the Hegelian teleological dialectic that shaped F. C. Baur's history of religion. Instead of conceiving rational dialectic as the natural process through which mind (*Geist*) develops a more universal and comprehensive self-understanding, Overbeck views the scientific project as a foreign interest that challenges existential valuations of life to rationally defend themselves. The interests of science and rationality, however, require justification on grounds that go beyond life. Just as Socratic dialectic signaled the end of the tragic view of life in Nietzsche's *Birth of Tragedy*, so here the rise of the scientific interest in knowledge *for its own sake* competes with and threatens the "vital forces" that originally give rise to valuations of life and the ideals they generated.[41] Overbeck writes, "If one surrenders the hitherto valid expression of the Christian-religious life to knowledge (*Wissen*), then it will succumb to it, since knowledge too must have arisen historically somehow and so must persist anytime that it comes at all into consideration. But knowledge can only persist when faith is too weak to stay by itself, and instead despairingly gives itself over to knowledge in order to concede to it or in order to be validated by it."[42]

Overbeck sees the appeal to reason and scientific knowledge in the arena of an existentially orienting ideal of life as a type of surrender and a sign of weakened vitality. To recall Bernard Williams's discussion of history and myth, science too has a history, and the rise of a scientific interest in the history of a *Lebensideal* signaled a threat to the naive self-confidence and strength of commitment that grounds such an ideal. Because the vitality and naive self-confidence of life was the source of its own justification, the introduction of new criteria of disinterested and value-neutral objectivity to serve as the new normative criteria fundamentally challenged life's claim to serve alone as the source of normativity. For this reason, "every theology, insofar as it brings faith in contact with knowledge, is in-itself and through this mixture *irreligious*, and no theology can arise where these foreign interests do not place themselves next to the religious."[43]

Interpreters have often captured Overbeck's dichotomy between faith (*Glauben*) and knowledge (*Wissen*) as one between "pre-conceptual" feelings, emotions, and sensations, and "concepts," thus seeing him through the lens of certain elements of German Romanticism and idealism. The lens has highlighted the theme of the exaltation of immediacy and feeling over "dead" abstraction and conceptual thought.[44] While helpful to some degree, this dichotomy misses the organic dimensions that Overbeck's conception of life introduces into the discussion of faith and knowledge. Instead of feeling against thought, Overbeck's fundamental dichotomy, like Nietzsche's, is between a natural impulse or animating drive that produces valuations of life "as a stem produces leaves" and an autonomous, disinterested interest in scientific knowledge *for its own sake*, disconnected from any such valuations. The conflict between life and science is thus also the conflict between religion and science, and Overbeck tells us that this conflict arises between two fundamentally different *aims* and *ideals* that arise out of different animating impulses and different sources of normative authority. A *Lebensideal* matters in the *vital* sense that it is something without which human life ceases to be livable, while the interest in science is something altogether different.

In light of this attempt to separate vital and scientific interests, Overbeck concludes that the task for theology and critical thinking with respect to religious valuations should not be to rationally justify or defend them against scientific knowledge. Instead, it should be simply to draw and police the lines of demarcation between these two distinct "magisteria."[45] These views explain Overbeck's curious position as a scholar who found no way for his work as a historian to serve the life that animated both Christian origins and the religious politics of his contemporaries. He found science to be crucially important, but on different grounds altogether from those that gave religions their connection to fundamental questions of human significance. As a historian, Overbeck could illuminate the history of Christianity and educate Christians about this history, but his expertise as a historian gave him no ability to speak for or against the ascetic *Lebensideal* that he took to be the core ethical message of Christianity. His worry was that cultivating scientific interest is simply a distraction, something altogether different, and thus potentially diminishing to vital interests in the value of life. Overbeck's Life-philosophy finally boils down to the claim that the project of putting *Wissen* in the service of an existential stance on the value of life ultimately introduces a foreign interest, aim, and set of normative criteria. It is for this reason that Overbeck believed that the rise of science and purely scientific interest weakened the vital interests out of which religions emerged.

This reconstruction of Overbeck's views of science and religion is a necessary prelude to understanding his appropriation of biological conceptions in later writings. In the *Christlichkeit*, Overbeck briefly turns to this problem by reflecting on the mediating efforts of Willibald Beyschlag (1823–1900), a contemporary theologian at the University of Halle in Germany, who exemplified a common theological response to Darwinism and evolutionary theory. Beyschlag attempted to mediate what was perceived to be the threat that evolutionary theory posed to theology by describing the "original cell" posited by evolutionary theorists in his day to account for the origin of life as a "miracle cell" (*Wunderzelle*).[46] Overbeck argued that this kind of attempt to marry theology and evolution was an acute example of precisely the sort of superficial mediation between science and religion that he hoped to expose. This effort captured neither the positivist aims of scientific knowledge to describe and explain reality in a value-free mode nor the religious interest in the value of life. While it aimed to marry the two, it really diminished both. As Overbeck pointedly remarks, "Of course for the religious person the Bible is not in the first place about natural science, but he also does not reflect upon the fact that there is none to be found in it."[47] As in the *Antrittsvorlesung*, the epistemic criteria and aims that constitute science were fundamentally different from the worldviews and ideals that constituted religions. Thus, Beyschlag's theological aim to defend the compatibility between Christianity and Darwin's tree of life, with its theoretical postulate of an "original cell" from which all of life descended, betrayed both a lack of historical consciousness and a crucial confusion about what religion and science are.

The *Kirchenlexicon*: Biology in History

Overbeck developed his theory of religion as a stance toward the value of life in his early *Christlichkeit*, and this conception was crucial for understanding his strident attack on the mediating aims of theology. Because of the dull reaction his polemical text met with, throughout the rest of his career as a purely scientific historian in the faculty of theology at Basel, Overbeck did not publish any further philosophical writings during his lifetime, preferring to confine his work to historical research on early Christianity. However, his historical research continued to build a case for his core genealogical thesis concerning the ascetic (de-)valuation of worldly life that was an animating impulse at the birth of the Christian movement. Later towering figures in theology and history, such as Ernst Troeltsch and Max Weber, would agree with Overbeck on the importance of asceticism—whether "inner worldly" or "otherworldly"—in the historical development of Christian ethics. And as we saw in the previous chapter, Overbeck's friend Nietzsche would come to see the birth of the ascetic ideal as the crucial event in the natural history of morality.

Overbeck's later historical studies defended the dynamic he had argued for between science and life through analysis of the changes Christianity underwent in its adaptation to and absorption of external and foreign interests of Greco-Roman culture. As Overbeck argued in the *Christlichkeit*, the attempt by early Christians to defend themselves, to adapt to an often hostile cultural environment, and to make a place for their religion in a world it initially thought would be coming to a swift end, forced Christianity to develop a scientific perspective on itself that ended up threatening the vitality of its original, ascetic valuation of life.[48] The conflict between science and religion in modern scientific culture was thus a continuation of the same dynamic struggle of the early Christian *Lebensideal* to assert itself in a world that had been educated in Greco-Roman philosophy and that forced it to defend itself on the grounds of reason rather than of its vital rejection of the final value of worldly pursuits. Overbeck saw evidence of this foreign intellectualism and scientific interest in the self-consciousness that early Christian thinkers developed about Christianity's own history, which began with Eusebius.[49] The birth of historical consciousness in early Christian historians indicated that the movement had turned, however slightly, from its original vital concern to a different set of interests and aims.

In the notes of Overbeck's planned *Kirchenlexicon*, he generalized this dynamic into a unique, biologically inflected theory of the developmental stages that religions go through in their confrontation with the time and history. The implication of Overbeck's biologism, as we will see, was very different from both the "affirmation of life" that Nietzsche drew from the biological world and the teleological view of history in Hegel and Baur. Overbeck writes in a key section entitled "Christianity (Time) in General":

> Until its total cessation, nothing is so old that it would no longer possess any part of its former youth. However, only a delusion can dream of a new layer of youth suddenly accumulating on organisms from somewhere else, and least of all such a new youth that would safeguard their transition to an enduring existence beyond themselves. What is eternal in us has always been in us, and did not first, in a historical moment of our life, become part of us. The experience humanity went through with Christianity cannot be asserted as a counter-example against this conception of history; [this conception] is rather confirmed through the [Christian] experience. In all seriousness, Christianity deserves the unusual historical interest which is given to it at most because it teaches humanity to grasp itself better and more easily as an organism of the general kind, one that originates, blossoms, and decays.[50]

Here, the fate of religion-in-history is seen as an example of the dynamics of ontogenetic development and the consequences of what natural history means for human cultural forms. The life-cycle processes of birth, youth, maturity, and old age become the underlying foundation for the dynamic of *Glauben* and *Wissen* that Overbeck observed in the history of Christianity as it traveled from original innocence and naiveté through experience in the world. To be-in-history, or to be historical, is to be subject to the dynamic of time, here understood through the biological cycles of youth, flourishing, and old age. As he writes elsewhere in the *Kirchenlexicon*: "[Historical] is not to be translated, but can it be circumscribed? How does it allow itself to be conveyed: subjected to time."[51]

The fate of Christianity in history was a struggle of an internal, vitally grounded principle against "foreign" interests, and this struggle signaled the subjection of such ideals of life to biological temporality. For Overbeck, Christianity itself came to have a history and to be-in-time in a way that it did not expect or intend. The anti-Hegelian elements of Overbeck's turn to organisms to describe historical development in these passages are clear. Nature and culture are not teleologically directed toward a fulfillment of themselves, their ideals, and their natures. Instead time and history are the processes through which the inevitable transience and finitude of distinctly biological processes assert themselves. The lesson of the life sciences and evolution, for Overbeck, was the transience of life as subjected to time. Overbeck's biologization of religious ideals transferred the temporal cycles of physical organisms to cultural forms, which also undergo a "struggle for life" to fulfill and realize themselves in a world that threatens them. In passages like the one above, Overbeck's pessimistic, Schopenhaurian biologism becomes a lens through which to interpret both the message of Darwinian evolution and the fate of Christianity in Western culture. Unlike his early "brother in arms," Nietzsche, Overbeck perceives the heroic affirmation of life not as a fundamental and characteristic drive of nature, but rather as a fleeting, "youthful" phase that must eventually fade.

In another entry, entitled "Religion (History)," Overbeck describes this eroding effect of biological temporality on a *Lebensideal*:

> Historical reflection on religion can only undermine its validity [*Geltung*]. For every religion belongs with respect to its origins among humans to a prehistorical world and can only find its end in the historical. In the historical, religion *can* become very old, but it can never escape the dangers of old age, that is, the circulation of coming into existence and dying away.... Historical life is also at the same time life-threatening. For what lives in front of our human gaze is not only 'good enough to perish' [*wert dass es zugrunde geht*]; rather it also does pass away, and that religion cannot extricate itself from this law of the world is the only thing that church history can and does teach about Christianity.[52]

These ominous characterizations of time and history allow a further glimpse into the general framework underlying Overbeck's appropriation of biological dynamics. Here, Overbeck introduces the term *validity* (*Geltung*)— a term that contemporary neo-Kantian philosophers like Hermann Lotze and later Rickert drew from the study of logic to capture the unique normative properties of values (*Werte*). Overbeck argues that the historical perspective "buries" the notion of validity because it shows values and ideals to be "temporary," subject to the conditions of time in general. Overbeck's radical historicism—his view that values cannot be considered objectively or a-historically valid—is grounded in his use of the biological life cycles of living things to understand historical development and change. The only teleology in the picture here is that of the inevitable phases of a life cycle and of finitude, not that of the broader process of evolution, history, or of a universal drive to life that underlies both.

Like Nietzsche, Overbeck uses categories of "vitality," "youthful/mature," "living/dead" in place of logical categories of "validity," "truth," and "falsity" to capture the claim that the normative force of values derives from the vital powers that create and erode them in history. In another passage, Overbeck addresses Nietzsche's understanding of philosophies and religions in terms of optimism versus pessimism in the judgment of the value of life. In an entry entitled "Pessimism," he writes,

> Dilemmas such as the one concerning a pessimistic or optimistic worldview, an egoistic or altruistic morality, are simply not to be permitted in scientific thinking. . . . For the world is what it is, beautiful and horrible at the same time, and rejecting the world or worshipping the world, as far as their human meanings are concerned, one is as justified as the other. It is not possible to marvel at the world enough, and no more possible to advocate for it, and the compatibility of its opposing properties eludes all comprehension. For that reason, one can do nothing better or higher with and in the world than to recognize it, to take it as it is and to let oneself be saturated [*durchdrungen*] by the indifference of one's value judgment [*Werturtheil*] on it.[53]

While Nietzsche saw in Darwinian nature a triumphal validation of the affirmation of life that he sought, Overbeck recommends the attitude of an ancient skeptic—to suspend judgment concerning the question of the value of life and, thereby, to achieve tranquility.

Despite Overbeck's contrary stance on the question of the value of life, these passages show that he too considered religions to be constituted by such basic judgments. Overbeck goes further in the *Kirchenlexicon* entry entitled "Pessimism" to echo Nietzsche's famous claim from the *Twilight of the Idols* quoted in the previous chapter that the value of life cannot be judged: "Just as the grounds [*Gründe*] of all value judgments [*Werturtheile*] lie outside the boundaries of humanity, so too do those concerning the question whether the world is good or bad. *All* religions prove only this, but prove nothing for the one side or the other."[54] Here, Overbeck's early analysis of the Christian ascetic view of life is linked to an analysis of value judgments in general. Overbeck sees optimism and pessimism, the valuation or devaluation of life, as judgments that arise out of life and not out of rational or scientific consideration of the world. The properly sober, scientific attitude toward the religious question was to remain indifferent to it—that is, to cease to view it as something of vital interest.

If modern theologians appear to Overbeck to be contemporary Don Quixotes, tiring themselves out on an impossible task of rationally and scientifically justifying a set of values whose source of justification and appeal lies outside of reason, what hope did Overbeck see for religion in his contemporary scientific culture? In two lengthy separate entries, Overbeck introduces the language of erotic life—of "love"—to capture these problems in a way that deserves full quotation:

> Human beings are not granted the ability to step into relationships that they would only have to defend, for none of these are able to withstand human critique. Luckily, however, human beings can love, and should love mother and father, their homeland, their siblings and friends, and their beloved, but these they should not defend. Love may not accomplish this as far as it would like, that is, through to the end—but this is also not a barrier for love. For love does not recognize any attack and is lost if it starts to worry itself. But it also cannot and does not need to do this, since it is based on a completely different foundation than the recognition of the objective, universal incontestability of its object that also generally holds for others, and thereby is not able to be pushed aside from its ground by any foreign attack. And so pessimism with respect to the world is tempered for people in a human way through love. If they only love the world, then all pessimism toward the world cannot harm them, that is, giving up defending the world cannot refute them in any special way. One ought to understand this not as a demand to 'give up defending the world,' but instead as the demand to give up taking the impossibility of defense more seriously than as something cheap [*ernster als billig*]. For such a general surrender of defense would also be a surrender of the dignity of all human existence, of all culture. Without criticism, we do not move forward, and of course this can turn us into false idealists. . . . We should also love our criticism, but still not think that it has unlimited value, that in our existence only our criticism and nothing better than this could have provided for us.[55]

Here, Overbeck announces himself as a resolute and yet still wholly ambivalent modern, committed to the life of criticism and scientific scholarship grounded in the principle of autonomy and freedom, and yet fully aware of it limits, its price, and the larger context of natural history and vital life in which the interest in scientific knowledge too had its source. Here, love is equated with life as the natural foundation of valuing that is neither objective, universally valid, nor rationally defensible. But it appears here too, as in Nietzsche, to be vital and redeeming. It is a force that protects, generates, sustains, and cares for ideals. Moreover, it sustains them against the disillusioning rationalist temptations to seek a universally valid, purely rational justification.

Overbeck characterizes modernity here as a condition of normative contest between rational and vital interest, in which historical consciousness and the epistemic values of science seem to make the question of the value of life appear strange. Overbeck concludes,

> Can one count religion too among the things that one loves without being able to defend them, that is, really loves? . . . What can calm us less than the nervous urge to defend religion that comes forward so strongly in the modern world? One feels the love that otherwise makes all defense so simple fading away. But does this not mean that the foundations of human love are beginning to break down with regard to religion? Indeed, does this happen because we are not and can no longer be the children that we must be, according to the Christian command, in order to be truly religious?[56]

Despite the ascetic and sober reserve of Overbeck's scholarly stance, there is an existential question at the heart of his project, and indeed his person, in which he has a stake—how can one, in a culture dominated by scientific interest, sustain a naive and resolute commitment to a set of values and ideals, of the sort that early Christianity allegedly had toward its ascetic *Lebensideal*? Can one sustain a sense of value in which a felt need for justification and critical defense, which turn out to be impossible, is absent? Overbeck's worry was that biological life, and its lack of "fixity" in a post-Darwinian world showed this naive attitude to be subject not only to historical but to biological time. The conflict between life and science, time and reason, the ideal and the real, was the sobering result of inquiry into the moral meaning of nature.

Conclusion

Overbeck saw the conceptions of the biological world emerging from the life sciences as vindications of his view that the human struggle to realize ideals is always a struggle *against* the temporal dynamics of nature and history. His own mixture of radicalism and reserve was a result of his skepticism that projects like Nietzsche's were either desirable or feasible. The notion that one could renew a second naïveté and new innocence that could revitalize European culture and supplant the valuations of life that had animated its theological heritage was simply not one that either scholars or new prophets could accomplish in an age of science. Modern Europe's so-called disenchantment and loss of naïveté in this sense was, ironically, the effect of the very scientific ideals that were its crowning achievement. Even Nietzsche's affirmation of life would appear to Overbeck to fare no better than the attempts by theologians like Beyschlag to marry evolution and the religious quest to find one's valuation of life reflected in the world. Religions, in Overbeck's view, did not need science or theology to interpret or justify them, and science did not need religion. However, for Overbeck as for Nietzsche, the concept of life created a bridge between the worlds of nature and the forms of human culture and value. Ultimately, the philosophical lesson that Overbeck drew from the Darwinian picture of nature was that religions, like life itself, were subject to finitude, temporality, and mortality.

One of the main criticisms that has been mounted against Overbeck both in his day and more recently is that he assumes a normative perspective on Christianity—one that took the values of "original" Christianity (*Urchristentum*) as what the normative Christian valuation of life was—that he could not justify on the grounds of his own theory of the vitalist sources of normativity.[57] Critics argue that Overbeck was not merely describing what early Christians wrote about the value of life or chronicling the historical emergence and fate of these values; he was making a claim about what view of life is normative for the Christian religion as a whole. In this sense, he too was making a properly theological, and not historical, claim that could not be justified on historical or vitalist grounds alone. This, indeed, is a version of the same criticism that we met with at the end of the last chapter in the guise of the so-called self-referential problem that Nietzsche faced. The claim, once again, is that both history and life could only be used to justify normative claims by appealing to grounds independent of history and life.

If taken as a criticism of Overbeck's view of what *Lebensideal* was normative for the teachings of Christianity, this point must be granted. There is no basis within Overbeck's general vitalist framework for the claim that otherworldly asceticism is the stance toward life that Christians should consider normative. However, this criticism misses some of the nuance in how Overbeck's genealogical project was related to his final evaluation of the Christian ascetic ideal. His focus on historical origins was aimed at exposing how different the valuation of life at the origins of Christianity was from the worldliness that he detected in the stance of modern theology. But Overbeck oddly expressed a favorable and even somewhat nostalgic view of the ascetic rejection of excessive worldliness that he studied in early Christian thinkers. He took the ascetic devaluation of life to be a sign of the wisdom of the Christian view of life; it was an honest acceptance of finitude. He felt that the worldliness of modern culture and theology failed to fully confront the finitude of life and history, and was, therefore, a sign of modern culture's own existential superficiality, typified in the writings of figures like D. F. Strauss. Overbeck's own criticism of the worldliness of modern culture, and of modern theology that had succumbed to it, was thus justified by his biologistic conception of life and time. The normativity Overbeck assigned to original Christianity had as much to do with his estimation of what he took to be its enduring wisdom in the face of the biological realities of life and time as it had to do with the fact that it was original. In this sense, Overbeck was indeed a theologian as much as a historian.

The relationship between Overbeck's genealogy and his criticism of contemporary theology is thus complex, and it rested on his notion of what religion was. Original Christianity spoke to vital interests, in Overbeck's estimation, because it was an honest response to the finitude of life. The ground of the normativity of asceticism in Christian ethics, for Overbeck, thus had to do with its ethical wisdom in facing life's realities, but this wisdom was, in his estimation, also the reason that it captured vital interest in the first place. If this is the case, then Overbeck does indeed have to appeal to a set of evaluative criteria other than life and vitality, such as wisdom, depth, authenticity, or honesty in the face of the difficulties of the human condition. But in Overbeck's view these were coupled in a way that is, in the end, not satisfactorily developed in his writings.

The overarching value of Overbeck's theology, and his science as well, was honesty. Yet, as Nietzsche's Life-philosophy remained in unresolved tension between eroding all sources of normativity and supplanting traditional sources with a notion of biological life, so too did Overbeck's. His own attitudes toward various existential ideals still held that their validity was transient and finite, grounded in fluctuating and finite vital drives. This would appear to undermine, or at least seriously qualify, his own commitment to the value of freedom and to the wisdom of ancient asceticism, which he found also to be at the basis of the scientific vocation. It is never clear whether Overbeck thought that the scientific valuation of knowledge had a different, more enduring and secure foundation than the transient, vitalist sources that gave rise to religions and sustained them. In his inaugural lecture, he grounded the value of science in the Protestant principle of freedom, and nowhere else did he argue that the scientific interest could be given anything more than such a genealogical, historical justification that relied on sources of the animating spirit of modern Western culture. Thus, we can surmise that the principle of freedom and the value of science could also only claim a vitalist origin, and so on this point Overbeck would stand firmly with Nietzsche.

This does not resolve the tension that remains in both of these thinkers between their own advocacy of vital wisdom and existential depth and their claims about the sources of values in youthful drives. Although his critics disagreed, Overbeck never viewed his inability to defend his commitment to the scientific vocation in a wholly scientific or rational manner as a failure. One consequence of Overbeck's Life-philosophy is, then, that there appear to be no rational or scientific grounds for defending valuations of life as genuinely normative. Yet, as we saw, Overbeck advised us not to take the impossibility of rational defense too seriously. His thought was above all marked by a tone of intellectual honesty about the limitations of science, scholarship, and rationality when it came to values, a stance that no doubt contains its own form of ascetic reserve. In his sober estimation, even the value of science could not be given any more secure foundation than the naive desires of life, but at least in his time he estimated these vital sources of science to be at the height of their youthful exuberance. The next chapters show how two towering voices in the next generation of readers of Darwin and Nietzsche posed further challenges for these conceptions of life and value, science and religion.

3

Georg Simmel: Evolution and the Self-Transcendence of Life

Georg Simmel's work on the combination of problems that animated both Nietzsche and Overbeck brought together many currents in the philosophical culture of fin-de-siècle Germany. A thinker whose broad interests are perhaps unparalleled in the history of philosophy, Simmel wrote on topics that ranged from the conceptual foundations of the emerging discipline of sociology to the philosophy of history, the foundations of ethics, fashion, adventure travel, leisure culture, art, urban life, economics, metaphysics, and religion. It seemed that every aspect of modern life piqued Simmel's dynamic and insightful philosophical gaze, from its only seemingly trivial details to its artistic, scientific, and metaphysical heights. Simmel's unique critical appropriation of Nietzsche and Darwin presents another form of Life-philosophy and a novel take on the lessons that thinkers from this period took from evolutionary biology. Simmel's concept of life took shape both in dialogue with Schopenhauer and Nietzsche and with contemporary neo-Kantian discussions about the radical difference between nature and normative validity (*Geltung*), the "is" and the "ought," that have been illustrated in the preceding chapters. Through a novel articulation of Nietzsche's thesis that the source of value is life, Simmel both challenged Nietzsche's answer to the question of the moral meaning of nature and recast the relationship between the constitutive values of scientific inquiry and the world's religions.

Simmel is primarily known in the history of philosophy as an antipositivist social theorist for whom there is no strict distinction between work in the social sciences and work in philosophy on the conceptual foundations of scientific investigation. As a philosopher, he was influenced by Nietzsche, Darwinian naturalism, Kant, Henri Bergson, and by his own neo-Kantian contemporaries, including Heinrich Rickert, with whom he exchanged many substantive letters on the problems of value theory that they both shared. Simmel understood himself as a neo-Kantian of sorts who wanted to expand what he took to be Kant's most important question, "How is nature possible?" to address the foundational question of the social sciences: "How is society possible?"[1] While Simmel's writings on the conceptual foundations of social science are extremely interesting and also connected to his broader Life-philosophy in important ways that I briefly mention, this chapter focuses on Simmel the theorist of value and the Life-philosopher. Like Nietzsche and Overbeck, Simmel turned to the concept of life to give a naturalistic account of the novel ethical and religious values that he observed around him in a highly diversified and rapidly changing modern culture. His notion of life was to answer both the genealogical (*quid facti*) question of the sources of these changing values and the transcendental (*quid juris*) question of the source of their normative validity.

Simmel began his productive career in the early 1890s with work that drew upon his studies in the fledgling school of social psychology founded in Germany by Moritz Lazarus (1824–1903) and Heymann Steinthal (1823–1899) and known as *Völkerpsychologie*. Frederick Beiser's recent and sweeping study of the German historicist tradition shows the pervasive influence of this early naturalistic school of social psychology on Simmel's philosophical work, even into his mature career.[2] Lazarus and Steinthal aimed to develop an empirical, scientific approach to what they called the *Volksgeist* (spirit of a people), a term they used to talk about shared affective and evaluative attitudes that underwrote communal norms, values, and shared identities. *Völkerpsychologie* was asserted as a naturalistic, empirical approach consistent with the sciences that could capture what Hegel meant to capture through his notion of "objective spirit" (*objektiver Geist*), but without its rationalism, intellectualism, and *a priori* transcendental metaphysics.[3] Simmel's pioneering studies in sociology and his work defining it as a scientific field built upon a naturalist and broadly empiricist foundation that focused instead on shared group instincts and affects that lay beneath the surface of national identities, shared ideals, rituals, and even the most everyday norms of social propriety.

However, even during his early engagement with this empirical school of sociology, Simmel had his eye on the great philosophical problems of his age. In 1892, he tackled the problem of historicism in *Die Probleme der Geschichtsphilosophie* (The problems of the philosophy of history). Simmel's sustained reflections on the theory of value began with his interventions in debates over *historicism*, a word that captures widespread worries mentioned in the introduction that greater awareness of cultural diversity and historical particularity undermined confidence in intrinsically valid ethical and epistemic norms. In the same year, Simmel also published his monumental work on the demarcation between the natural and human sciences, entitled *Einleitung in die Moralwissenschaft* (An introduction to moral science).[4] Even Simmel's dissertation on Kant's conception of nature betrayed the philosophical aims that would fuel his mature, biologically inspired Life-philosophy later in his career.

Simmel's intellectual development beyond his early engagement with *Völkerpsychologie* has been charted differently by scholars, but a general agreement on distinct early, middle, and late phases is common. One way of characterizing the phases of Simmel's thought is through the schema of an early, positivistic phase, in which Simmel utilized the broadly Darwinian and naturalist theory of culture in *Völkerpsychologie*, followed by a middle, neo-Kantian and "critical" phase, and finally a late, "metaphysical" stage of Life-philosophy.[5] Klaus Christian Köhnke's *Der Junge Simmel*, a definitive study of Simmel's early evolutionary and positivistic phase, identifies Simmel's turn to a neo-Kantian theory of value in his influential work of 1900, *Die Philosophie des Geldes* (*The Philosophy of Money*). Köhnke argues that Simmel's early, precritical phase was Darwinian, characterized by a naturalistic framework that derived of cultural values and the "objective spirit" of cultural groups from the basis of given and dynamic "feelings of value" (*Wertgefühle*) that originate in instincts and given bodily drives. In contrast to this, his postcritical neo-Kantian work stresses the stability and fixity of ideal cultural forms of life over against the dynamic flux of vital drives.[6] This description is helpful for getting at the way in which the neo-Kantian focus on *Wertphilosophie* (philosophy of value) pushed Simmel to augment his early Darwinism and to give more autonomy to the sphere of the ideal in relation to its origins in natural drives. However, this chapter challenges this clear-cut periodization of Simmel's work by arguing that Simmel's late, metaphysical Life-philosophy is not a departure from his earlier Darwinism and later neo-Kantian concerns with value theory. Instead, the later phases of Simmel's thought build upon his early Darwinian and naturalist theory of cultural values with the aim of making it more defensible against the neo-Kantian arguments for the autonomy of nature and normativity, the "is" and the "ought." Simmel's Life-philosophy moved from naturalism through transcendental idealism and on to full-blooded metaphysics because it attempted to resolve the conflict between the "is" and the "ought" while maintaining the insight of his early Darwinian phase that the sources of value in lie in the pre-rational, instinctual characteristics of evolving life.

As with Nietzsche and Overbeck, Simmel's eclectic style, the volume of his writing, and the development of his thinking throughout his career pose serious challenges for any attempt to summarize his thought cleanly. As an example of the wide-ranging nature of Simmel's publications, between just the years 1905 and 1907 he published two extensive essays, "The Philosophy of Fashion" (1905) and "Religion" (1906); a monograph on *Kant and Goethe* (1906); and finally a lengthy monograph on *Schopenhauer and Nietzsche* (1907) that he saw as a complement to his work on Kant.[7] The challenge of Simmel's diverse interests and voluminous output, however, also affords a unique opportunity once again to show a thinker for whom the concept of life unified an otherwise widely diverse array of topics. In order to make the systematic connections in his thought clear, I do not proceed chronologically at first. I begin by analyzing Simmel's interpretation of Nietzsche in his post-1900, post-positivist and neo-Kantian phase. I then return to Simmel's earlier *The Philosophy of Money*, which addresses neo-Kantian criticisms of naturalism and extends his early Darwinian theory of value. Then I examine the central themes of his late metaphysical *opus* in Life-philosophy from the last years of his life, *Lebensanschauung* (The View of Life). Finally, I turn to Simmel's theory of religion to show how his conception of life and value fed into a unique understanding of the conflict between religion and science.

Simmel's Nietzsche

The shadow of Nietzsche's approach to values, religion, and life is everywhere in Simmel's work. From Simmel's early studies in *Völkerpsychologie* to his monumental, two-volume, naturalistic critique of the foundations of ethics and the influence of Darwin on his own later, so-called pragmatist and evolutionary epistemology, Simmel could be said to have been primed to be receptive to the still contentious legacy of the naturalistic iconoclast. The *Introduction to the Moral Sciences* from 1892/93 argues in Nietzschean style for the affective origins of ethical values, for the impossibility of justifying values on strictly rational grounds, for the importance of history and psychology for explaining the origins of religious ideals, and, finally, it champions an "individual law" against the universalism of Kant's ethics and its categorical moral imperative.[8] It is not surprising, then, that Simmel was one of the first scholars to take Nietzsche seriously as a philosopher. He published short essays in the 1890s on Nietzsche and packed lecture theaters in Berlin in 1902 for a series of lectures on the philosopher, whose legacy was still very much a matter of contest.[9] In 1907, Simmel published a monograph, *Schopenhauer and Nietzsche*, which is still widely recognized in Nietzsche scholarship as offering the first non-metaphysical reading of Nietzsche's famous doctrine of the "eternal recurrence of the same," seeing it as a thought-experiment designed to give gravity to one's actions and decisions in the present, instead of as a serious theory of the cycles of the cosmos. This book was pioneering in its attention to the importance of Schopenhauer to Nietzsche and for its early recognition of Darwin as crucial figure for understanding Nietzsche's project.[10]

Above all, Simmel regarded Nietzsche as a peculiarly modern thinker, the prophet of a new modern *Weltanschauung*, but one whose roots were in the great social and economic transformations of nineteenth-century European society. Simmel took it as a central goal of his life's work to bring the anthropological lens that was then being applied to so-called primitive cultures in the school of *Völkerpsychologie* to the study of modern life, its changing sensibilities and values, and the concrete sociohistorical conditions of its emergence and proliferation. Nietzsche became a focal figure because he, more than any other, embodied the "spirit" of this new, modern age. However, for all of their affinities, Simmel was also a sharp critic of Nietzsche, and he answered the questions Nietzsche raised in different ways. As an early champion, in his 1892 *Einleitung in die Moralwissenschaften* (Introduction to the moral sciences), of the relativism, historicism, and psychologism that the neo-Kantians so strenuously fought against, Simmel was disposed to embrace Nietzsche as his own philosophical "brother in arms" against the rationalism and antinaturalism of neo-Kantian *Wertphilosophie* (Value-philosophy). Simmel's attempt to make improvements on Nietzsche stemmed from the burden of making Nietzsche accessible and philosophically compelling to his academic peers, such as Rickert, the subject of the next chapter. This largely hostile audience was steeped in the "late-Idealist" appropriations of Kantian and Hegelian philosophy and its continued emphasis on the autonomy of *Geist* from *Natur*.[11]

Simmel's 1907 monograph on *Schopenhauer and Nietzsche* reconstructed the core themes of both of these philosophical renegades of the nineteenth century and demanded recognition of their philosophical importance. They belonged together not simply because of Nietzsche's explicit debt to Schopenhauer, but because they constituted a distinct and coherent intellectual lineage that challenged the Hegelian aggrandizement of an autonomous and self-determining reason and emphasized the nonrational dynamics of will and affect as more fundamental to the development of human culture. Simmel's monograph was one of the first scholarly studies of the movement of Life-philosophy that he would later write himself into. Indeed, Simmel already in 1907 argued that Nietzsche had "Darwinized" Schopenhauer by linking the principle of the ever-striving will underlying the manifest, phenomenal world to Darwin's idea of an endlessly changing and dynamic flux of organic forms over geological time.[12] While Simmel's analysis of Nietzsche's

Darwinism is similar to that offered in the first chapter, it also differs in crucial respects that I identify below.

Simmel righty pointed out that Schopenhauer and Nietzsche set themselves apart from Kant and the post-Kantian German Idealists through their anti-intellectualism, or the priority they accorded to affect and desire over intellect and reason in the formation of values. For Schopenhauer and Nietzsche, the most formative forces shaping human experience were desire and drive. The insatiable appetite of erotic drives for satisfaction at all levels was the creative principle behind the generation and destruction of both biological and cultural forms. Indeed, this dynamic desiring and striving created the metaphysical link between human psychology and the formative processes underlying biological life as a whole. This meant that the psychology of valuing, of positing values that were taken to be normative and striving to realize them, became the key link between biological processes and the ideals that constituted ethical, scientific, and religious life. For Simmel, it was Schopenhauer's and Nietzsche's adept and subtle psychological insights into human *eros*—their moral psychological insights into the complex motivations behind human beliefs and behaviors—that finally established their philosophical importance over against the rationalist tradition. Simmel allied himself with Nietzsche's claim that the psychology of valuing, and not logic or the self-critique of reason, was the new path toward the foundations of knowledge, ethics, and religion.

Simmel argued that it was necessary to see Schopenhauer and Nietzsche as two opposing sides of a single coin.[13] For Schopenhauer, "the will is the substance of our subjective life, just as and because the absolute nature of being is a restless urge, an always going-beyond-itself, which, precisely because it is the creative source of all things, is condemned to eternal dissatisfaction."[14] Simmel regards Schopenhauer as a post-Christian thinker who encapsulated the sensibility of a European modernity for which the Christian view of life had lost its hold over the existential imagination, but the spiritual hungers that had sustained this view continued to linger. Simmel writes further, "This longing [for an end-goal] is the inheritance of Christianity," which "left behind the need for a *definitivum* of the movement of life that continued on as an empty longing after a goal that had become unreachable."[15] The central question raised by Schopenhauer's understanding of the formative impact of Christianity on the erotic life of European culture was to what extent these "longings" and metaphysical "needs" for an absolute, a point of stability and stasis, were essential to human nature—or were they rather the lingering traces of a merely temporary cultural form, a value-laden clothing for life that would also be subject to Darwinian transformation. In Simmel's view, Schopenhauer built his metaphysical system around the experience of existential dissatisfaction, an experience of frustrated desire. This, then, was the psychic manifestation of a restless *eros* that was the driving force of the world as a whole. But whereas Schopenhauer recognized only one adequate response to the inevitable dissatisfaction of the restless will—namely, extinguishing the erotic desire for more life—the philosophical labors of his heir Nietzsche were directed precisely at the opposite: life itself in its restless movement was to become the source of its own satisfaction.

Simmel saw the starting point of Nietzsche's philosophy precisely as Schopenhauer's sense of a world robbed of a transcendent *telos*, but he identified Darwin as the intermediate influence that offered Nietzsche a less pessimistic interpretation of nature. Simmel credited Darwin for supplying Nietzsche with a new conception of the *teleological* character of life that allowed him to come to a different conclusion about its value. According to Simmel, Nietzsche conceived of life and its evolution in a progressive sense, arguing that it tended always toward enhancement, growth, expansion, and increase in power. Simmel writes, "Through this drive which immediately lies within [life] and the guarantee of enhancement, enrichment, and consummation of value [*Wertvollendung*], *life can itself become the goal of life*, and thereby the question of a final goal that is to lie beyond its pure and naturally running process is superseded."[16] This interpretation of Darwin's influence on Nietzsche is subtly different from the view defended in the first chapter. There, I argued that, in Nietzsche's view, life is able to serve as the basis of its own value because of its teleological character as self-appetitive and, therefore, also naturally self-affirming. Simmel recognized this as well, but argued that Nietzsche conceived of the entire process of evolution—not just its products—as *teleological* in the same sense. That is, evolution was a process that pressed toward the enhancement of nature and an ever greater consummation of the creative powers inherent in nature.

Simmel rightly argued that the lesson Nietzsche drew from Darwinism was that nature is not indifferent, value-free, or value-neutral. Darwin offered a nature that itself dynamically pushed toward values and, thereby, was *constitutive* of values. Most importantly, natural life was constitutive of the value of *itself*. Simmel argued that Nietzsche's marriage of teleology and Darwinism offered a solution to the problem of the "is" and the "ought," or, in the philosophical terminology of late nineteenth-century Germany, *Sein* and *Sollen*, which had become by the time of Simmel's writing the most pressing problem of the Southwest, neo-Kantian school in the German academy through the work of Hermann Lotze, Wilhelm Windelband, and his student, Heinrich Rickert. The widespread philosophical problem had largely been fueled by debates over the scope of the natural sciences in relation to philosophy and over the legacy of Kant's epistemology. Simmel rightly showed that Nietzsche's Darwinian notion of life fused fact and value: the idea that nature has a meaning *for life* was itself an assertion that one could draw valid values (*Sollen*) from a doctrine of "being" (*Sein*) because being inherently strove toward value.

For Simmel, Nietzsche's use of Darwinian life to fuse fact and value, biology and philosophy, was the most significant aspect of his project. Nonetheless, Simmel turned around and criticized Nietzsche's biologism in ways that mirrored the attacks on psychologism in the theory of value that were being forcefully made by neo-Kantian contemporaries like Rickert. These attacks insisted that psychological events and norms of rational thinking occupy two distinct and logically incompatible spheres. The charge of psychologism related precisely to the necessity of distinguishing between normative and descriptive/empirical claims. R. Lanier Anderson helpfully characterized the difference between normative and descriptive claims in terms of the "direction of fit" between concepts and the world.[17] Descriptive claims fit themselves to the regularities of the observed world, but norms act as criteria that the observed world is to fit itself to. If an observed event breaks what is thought to be a law of nature, then the law is not valid and must be reformulated; but if something in the world does not live up to a norm, then something is faulty with the world and not with the norm.

Simmel argued in his book on Nietzsche that the Darwinian principle of life spuriously elevated the purely natural course of events in the biological world into a norm of conduct and value and so surreptitiously snuck values into the blind course of nature. For Simmel, then, Nietzsche's naturalistic description of life and his aristocratic ethic of heroic life-affirmation were caught in a vicious circle. His analysis on this crucial point deserves to be quoted in full:

> The concept of life, to which evolution gave a new meaning, appeared to have achieved what was sought after from all sides: to derive logically the content and meaning of the ought [*des Sollens*] from a given, identifiable reality. That, of course, was the enormous difficulty of all ethics and doctrines of value: that is, that the most general, ethically necessary and valuable principles could not be derived from the demonstrable and real, and thereby appeared to be sacrificed to individual arbitrariness and purely personal opinion [*Überzeugung*]—a consequence that almost every metaphysics believed it could avoid by secretly inserting [*hineingeheimnissen*] the good and the ought into the "actual," the really real. . . . Nietzsche's selection of that which he rec-

ognizes to be most valuable within real life is not inscribed in the structure of this reality, but instead can only come from a feeling of value [*Wertgefühl*] that is independent of it. And only an optimistic, enthusiastic "faith" [*Glaube*] in "life," which shares total unverifiability with Schopenhauer's pessimism, can regard these values, whose constitution flows from completely different sources, as the nerve of life itself, as the factors of its actual enhancement.[18]

Simmel, like Overbeck, argued that Nietzsche's affirmation of life was simply his own "faith" (*Glaube*), it was itself the product of Nietzsche's own "feelings of value" (*Wertgefühle*) and not an insight into the nature of reality itself. Simmel thus exploited the self-referential problem referred to at the end of chapter 1, which was responsible for the unresolved tension between Nietzsche's championing of the value of life and his conception of the nonrational sources of normativity. If values are the products of drives and affects, even ones shared by other living things, then the affirmation of life could only claim validity as a subjective feeling and not as a criterion of human flourishing. To put it crudely, the dynamics of evolving life are what they are, but even to pose the question of the value of life was to make an unjustifiable leap from Darwinian science into another sphere altogether.

Life and Value In *The Philosophy of Money*

Simmel's analysis of the debate over pessimism and optimism that divided Schopenhauer, Nietzsche, and Overbeck raised a core problem that recurred throughout his middle and later writings. It may seem odd after reading of Simmel's deep engagement with figures like Nietzsche and Schopenhauer to turn now to a work that solidified his status in the history of modern social theory next to the likes of Durkheim, Weber, and Marx. Yet the connection between Simmel's interest in social theory and his interest in Nietzsche and neo-Kantianism is not accidental. What Simmel wrote of Nietzsche in his 1907 book indicates this link between political economy and life:

> [Nietzsche's] noble person does not ask, "What does it cost?" For that reason, the style of the noble life is so completely contrary to that of the money economy, in which the value of things becomes more and more identified with its price. [Hippolyte] Taine tells of the highly wasteful aristocracy of the *ancien régime*, where it would have counted as a symptom of nobility that one laid absolutely no value in money.... The deep aversion of Nietzsche to all specific appearances of the money economy has to be traced back to the fundamental antinomy that exists between the ways of valuing of the money economy and those of an aristocratic mode of valuation.[19]

In good historicist fashion, Simmel contextualizes Nietzsche's aristocratic ethics by explaining it as a reaction against the materialistic, bourgeois values that were coming to dominate the modern industrial money economy. Simmel's monumental and characteristically sprawling work *The Philosophy of Money* from 1900, for which he is most well-known, is a study of the effects of the developed money economy of Western industrializing and urbanizing nations on new individual and collective psychological conditions that gave rise to new values. It exceeds the limits of this study to do full justice to this nuanced, provocative, and wide-ranging work, but I use it in the following to identify important aspects of Simmel's entanglement with Darwinism, neo-Kantianism, and the theory of value.

Simmel's project in *The Philosophy of Money* is complex, but he hints at his unique guiding question by claiming that its aim is not to provide an "economic" theory of money, but a "philosophical" one.[20] What Simmel means by this is that his aim is to investigate the connection between the economic function of money in the modern money economy—which he considers to be the most significant factor shaping modern sensibilities—and the ethical values and styles of life that proliferate in this ever-expanding new social form. In short, his goal was to study the effects of economic conditions on ethical and even spiritual ideals related to the aims of human life in general. As Simmel writes, "In this constellation of problems, money is simply a means, a material or example for the presentation of relations that exist between the most superficial, 'realistic,' and fortuitous phenomena and the most idealized powers of existence, the deepest currents of individual life and history. The meaning and purpose of the whole project is just this: to derive a guideline from the surface of economic events that leads to the ultimate values and significance of all that is human."[21] Simmel's ambition was massive but his thesis was simple: the money economy leads humans to equate value with monetary price, and it thereby supplants the decidedly ethical concept of "intrinsic" value with the economic concept of "instrumental" value.[22]

The goal of Simmel's ambitious study expresses the more general sociological-psychological perspective already visited above. Philosophical, religious, and ethical valuations are understood as local, contextual, and historically variable; they are both shaped by and responsive to the psychological effects of specific sociohistorical and economic formations. This relativizing, historicizing, and naturalizing move is the foundation of Simmel's philosophy of value, which he began in his early writings on the moral sciences and developed in *The Philosophy of Money* before making it the basis of his interpretation of Nietzsche and Darwin. What makes this work distinctive is its strikingly original thesis that the philosophical position of what he calls "relativism," the turn to evolutionary biology as a foundation of value, and religious and ethical values of "individuality," "creativity," and "freedom" in modern culture are all fundamentally interconnected features of a distinct and historically novel "form of life," shaped by the social conditions of the modern money economy.

These concerns no doubt bring to mind another nineteenth-century "master of suspicion" whose cultural impact was far greater than Nietzsche's—Karl Marx. Simmel's *Philosophy of Money* was an explicit response to the Marxist tradition and its economic reduction of social and ethical values to modes of production. He writes that his philosophical study of money aimed "to construct a new level beneath historical materialism [i.e.,—of the Marxist tradition] such that the explanatory value of incorporating economic life into the causes of intellectual [*geistig*] culture is preserved, while these economic forms themselves are recognized to be the products of deeper valuations and currents with psychological, even metaphysical, presuppositions."[23] Simmel thus wished to correct Marx's materialist reduction of cultural values and ideals (the so-called ideological superstructure) to material and economic factors relating to production alone by identifying a more primordial source of economic values themselves. Later in Simmel's life, he would return to Nietzsche and Darwin to claim that this source was the vital current of life. The central problem with the Marxist reductionism in this work, Simmel complained, was that it took economic value for granted as a given fact and thereby left it underived.

Simmel's *Philosophy of Money* thus countered the prevailing reductionism of historical materialism by raising concerns over the relation between values and being that resonated throughout Nietzschean and neo-Kantian philosophy. Simmel's neo-Kantian investigation into the conditions of the possibility of economic value was not a genealogical investigation into its *natural-historical* origins in the evolutionary history of the human species, but into the rise of values out of basic, pre-rational drives. As Simmel described it, a philosophy of money had to go beyond the economic science of money to find "preconditions that, situated in the constitution of the psyche, in social relations, and in the logical structure of reality and values, give money its meaning and practical position."[24] Simmel's study is not an attempt to answer the transcendental, *quid juris* question of what right money has to the value ascribed to it; rather, it is an inquiry into how money acquires value in the first place. This can only occur, he claims, by relating money to fundamental human drives and desires. In this sense, he saw economics as a branch of the more fundamental sciences of biology and psychology.

Simmel saw the ambitious project of applying a neo-Kantian set of concerns to the economy as practicable, even necessary, because of his underlying conception of the relationship between *Sein* and *Sollen*. He writes,

> That objects, thoughts, and events have value can never be read off of their mere natural existence and content; and their ranking according to value diverges widely from their natural ordering. Countless times nature destroys that which from, the standpoint of its value, calls for longer duration, and conserves worthless things that take the place of the valuable ones. . . . Value first comes to the completely determined objective being like light and shade, which are not inherent in it but come from a different source. However, the misunderstanding should be avoided that the formation of value concepts, a psychological fact, is distinct from the natural process. A superhuman mind that could understand completely everything that happens in the world according to natural laws, would also comprehend the fact that people have concepts of value. But these would have no meaning or validity beyond their psychological existence for a being that conceived them purely theoretically. What is denied nature as a mechanical causal system is just the matter-of-fact content and *meaning* of value concepts, whereas the psychic events that make this value-content a part of our consciousness by implication belong to nature.[25]

This key passage gives us Simmel's view of the ontology of value. First, he argues that value, like light and color, is not an intrinsic, primary quality inherent in objects themselves, but instead arises only through a *relation*, in particular a relation to a valuer, some being that carries out the act of valuing something. Second, Simmel affirms that valuing is a psychological fact about human beings; like Nietzsche, he sees valuing as something that humans necessarily do as part of their belonging to the natural order. Although current scientific methods might be limited in their ability to actually trace the complex, so-called proximate causes of psychological acts of valuing, nonetheless Simmel claims that a superhuman mind that could achieve a full description of nature would be able to. However, Simmel finally and intriguingly argues that even though such a naturalistic explanation of value is in principle possible, there is a limitation even to the ability of such a superhuman "view from nowhere" to comprehend value.[26] This is because the psychological act of valuing is required in order to understand the meaning or content of a value, and to even raise the question of its normative validity. By claiming that a superhuman mind that viewed value "purely theoretically" would not be able to comprehend the question of the validity of values, Simmel contends that what it means to claim that something has value can only be comprehended by a valuer and not, as it were, from the outside. Just as color concepts might not mean anything to one without sight, so the absence of a practical standpoint, grounded in desire or appetitive response to the world, would make the concept of value unintelligible.

Simmel argues on the basis of this picture that humans have a peculiar relationship to the problem of value that arises from inhabiting these two distinct standpoints on the world. On the one hand, humans occupy a limited version of the theoretical perspective of the hypothetical "superhuman mind," forming a disinterested picture of the world as a sum total of causal relations. Forming this picture, moreover, is the project of science. On the other hand, humans inhabit the practical perspective of a valuer in their appetitive and affective relationship to the world. He argues that the peculiar condition of the human being as a thinking agent, both an active valuer and a knower, generates the paradoxical incommensurability between the "is" and the "ought," *Sein* and *Sollen*. It is a psychological fact for Simmel that humans must posit values as they actively and appetitively interact with the world. The problem of whether or not such values could be thought to be *valid* was precisely the problem of which of these two standpoints, the theoretical or the practical, was to have priority in determining what the world is *really* like.

The distinctive mark of Simmel's theory of value is the claim that this aporia of the unity of *Sein* and *Sollen* was generated by the conflict between practical and theoretical standpoints on the world and so could not be resolved from within the human perspective. The stance of disengaged theoretical knowing might explain psychological acts of valuing with reference to causal interactions in nature, but it could not identify such psychological events as acts *of valuing* or assess the content of these values without occupying a practical, engaged stance. Humans were thus stuck between the conflicting standpoints of science and life. Simmel goes on to defend a solution to this incommensurability between the conception of causal relations and the conception of values that arises out of the human standpoint by invoking Spinoza's monistic response to the Cartesian and dualist distinction between extension and thought, the physical and the mental. For Simmel, being and value are two distinct and partial perspectives on a unified underlying reality that is, ultimately, distinct from them both.[27]

The connection between Simmel's prolegomenon to the theory of value and his turn to the concrete phenomenon of money is crucial for grasping some of the basic assumptions that would later inform his Life-philosophy. Unlike the superhuman mind of his hypothetical thought-experiment, Simmel's goal in *The Philosophy of Money* is not the theoretical goal of showing how psychological acts of valuing are caused in nature, nor is it the metaphysical goal of characterizing the underlying monistic unity of being and value. Simmel's goal is rather to provide a genealogical theory of the *experi-*

ence or phenomenology of objective value, an account of how the phenomenal appearance of living in a world with an *autonomous* order of values distinct from subjective acts of valuing has come about for embodied subjects. He thus wishes to explain the origins of the phenomenology of the practical standpoint itself —how it is that the experience of an independently existing, shared world infused with valid values arises out of social interaction between discreet individuals.

Simmel approaches normative validity not as a property of values, but as a phenomenal fact about how individuals experience value in the world. He writes,

> In whatever empirical or transcendental sense one can speak of the difference between "things" and subjects, value is never a "quality" of the objects, but a judgment upon them, which remains inherent in the subject. Still, neither the deeper meaning and content of the concept of value, nor its significance for the mental life of the individual, nor the practical social events and forms based upon it, can sufficiently be understood by referring value to the "subject." The way to a comprehension of value lies in a region in which that subjectivity is only preliminary and actually not very essential.[28]

Even though values must be conceived, ultimately, as arising out of the interaction between individuals and their "subjective," psychological acts of valuing, Simmel recognizes that philosophers bear the burden of understanding how values nevertheless assume significance for individual agency, social life, and culture that is no longer merely subjective. The theory of the genesis of objective value that Simmel offers is quite simple: in order to experience an object as possessing value *in-itself*, and not merely for me, there must be an increasing distance between the immediately felt subjective need or desire for the object and the satisfaction of that desire. In other words, in order for something to be considered *objectively* valuable, it cannot be directly and immediately enjoyed.

Simmel argues for this genealogy of value by depicting an inverse relationship between the intensity of immediately felt subjective emotions and ability to represent an objective world. Simmel writes, "Everywhere the weakening of the affects, that is, the unconditional surrender of the Ego to its momentary feelings, is correlated with the objectification of representations, with their appearance in a form of existence that stands over against us. . . . The tranquilization of the passions, and the representation of the objective world as existing and significant, are two sides of one and the same basic process."[29]

This is where natural and social history begin to play important roles in Simmel's philosophy of value. Simmel fits his account of the development of the modern money economy into a naturalistic, Darwinian story of the evolution of human cognitive capacities. The less humans were ruled over by momentary instincts and passions, the more these feelings became weakened, and the more it became cognitively possible to represent abstract, objective qualities of autonomously existing things. Only when want and satisfaction, desire and enjoyment, did not coincide did it become possible to represent values as valid independently of one's desire.[30] The concept of objective validity thus arose out of the loss of immediacy, and this occurred through a unique evolutionary process whereby human cognitive abilities developed through social interaction. Social evolution gave the human mind the ability to represent the world as an autonomously existing, objectively valuable sphere in which to act. In relation to economic values, Simmel concludes, "One can say, therefore, that the value of an object does indeed depend on the demand for it, but on a demand that is no longer instinctive."[31]

Money is the prime example that Simmel uses to transition from the subjective origins of valuing and desire to the social fact of an objective world of independent values. His theory of money takes the process of exchange to be crucial for creating the necessary distance between objects and their enjoyment. Exchange, for Simmel, is the process through which the natural, given aspects of subjective life, such as biological needs and desires, acquire a properly *economic* meaning. It is also the process that accounts for the function of money as an abstract medium, a neutral metric of value that no longer is tied to the distinct qualities of the objects that are exchanged and that give them value for humans in the first place. Through exchange, objects gain increasing distance from the subjective life that is the original ground of their value, and these objects then come to confront one another as though their value was a property inherent to them. Simmel writes, "We cloak economic objects with a quantity of value as if it were an inherent quality, and then hand them over to the process of exchange, to an objective mechanism determined by those quantities, to a confrontation between impersonal values—from which they return multiplied and more enjoyable to the final purpose, which was their starting point: subjective experience. With this the formation of value begins and the direction through which the economy realizes itself is established, whose consequences represent the meaning of money."[32]

The process of exchange embodies the fact that objects are valuable not merely for an individual but for other subjects. Money becomes the outward, external sign of the psychological experience of objective value, but this is a value stripped of personal significance. As Simmel writes, "The equation objectivity = validity for subjects in general, finds its clearest justification in the economic form of value."[33]

Simmel's Metaphysics of Life

Though many of the basic elements of Simmel's naturalistic, evolutionary approach to value are already hinted at in *The Philosophy of Money* and in Simmel's work on Nietzsche, Simmel only devoted himself to the full systematic articulation of his conception of life in his final years. Simmel's last book, *Lebensanschauung* (*The View of Life*), which he worked on for four years up to the last days of his life in September 1918, was the closest he came to a synthesis and systematic exposition of the ideas contained in his voluminous sociological and philosophical essays on modern culture. This late work is above all a meditation on and attempted resolution of the underlying unity of being (*Sein*) and the normative ought (*Sollen*) that featured in his early reflections on Nietzsche and economic value. Simmel revived Nietzsche's concept of life to for very much the same purpose: to unify the underlying dynamics of nature and the phenomenal world of objectively valid values within which humans think, act, know, and experience the world as a meaningful whole. Furthermore, this concept allowed him to connect his theory of value to art, religion, science, law, morality, and even logic.

Simmel's intuition about the common origin of disparate domains of modern culture stemmed largely from his observation that they all shared an important feature—they were constituted by values that made an objective claim on human agency and thought. As in Nietzsche's project to show the origin of science in the natural drive of "life," Simmel's goal was to show that the values aimed at in scientific investigation and in philosophy too originated out of the same value-constituting reality as aesthetic, religious, moral, and even economic value (hence developing the basic insight of his *Philosophy of Money* that economic value is merely a species of the genus of value more generally). The concept of life, for Simmel, was a comprehensive, unifying principle, an identification of the One in the Many and of the underlying unity in the diversity of both nature and culture. The search for this final unity led Simmel in his late work to adopt self-consciously metaphysical aims.

The basic strategy of *The View of Life* is continuous with *The Philosophy of Money* and with Simmel's early Darwinian and sociopsychological framework. His articulation of the concept of life is driven by the genealogical question of how apparently autonomous and objective orders of valid values (*Sollen*) arise out of underlying natural processes (*Sein*). Simmel defines culture through the concept of "ideal worlds" (*Ideale Welten*), and *life* is his term for the natural processes operating in the whole of nature that also generate these worlds.[34] Simmel uses the concept of an ideal world throughout this work to characterize domains constituted by "mental" (*geistig*) activity, which include art, science, morality, religion, and law.[35] These ideal worlds are constituted by norms and values that become enshrined in the institutions of social life. In *The View of Life*, Simmel acknowledges his debt to Nietzsche as a precursor to his genetic project of searching for the origins of ideal worlds in life. However, against Nietzsche, Simmel argued that the defining characteristic of these ideal worlds is that they have achieved autonomy from the life out of which they emerged, and this fact about the autonomy of values was something that Nietzsche's Life-philosophy, in his view, failed to capture.

Simmel's formula for life is simple, but abstract enough that it requires a bit of unpacking. "Life," he claims, "has two mutually complementary definitions: it is *more life*, and it is *more-than-life*."[36] The simplicity and abstractness of this formula belie the heavy theoretical work that it does. It is meant to account for the creative dynamic that underlies changing organic forms, and cultural forms that arise with human intellect and abstract thought. Simmel's notion of life as both "more life" *and* "more-than-life" grants ideal worlds autonomy, but merely *relative* autonomy, from the natural drives that produce them.[37] Ideal worlds are not moments in a vital Heraclitean flux driven by life's "lust after itself;" they are abiding forms of relative stability and fixity, dams that momentarily block, collect, and quarantine the creative current of life. For this reason, Simmel dismisses the idea that the process of life is continuous creation of more of itself (as it was for Nietzsche). If so, the possibility of ideal worlds of culture whose values are not reducible to biological needs and desires would not make sense. The crucial puzzle was how to account for the fact that the guiding values that constituted ideal worlds—those forms of art, religion, economy, and even science—were experienced as intrinsically valuable *for their own sake*.

The very existence of stable cultural forms constituted by guiding values, for Simmel, demands that life be understood as a self-transcending and creative process, one that produces its opposite and tends toward something that is "more-than-life." He writes,

> [Life] transcends itself when it is not only more life, but more-than-life. This is always the case when we call ourselves creative—not only in the specific sense of a rare, individual power, but in the sense that is obvious for every representing of something, that representation produces a content that has a meaning of its own, a logical coherence, a certain validity or stability independent of its being produced and carried by life. This autonomy of the created thing speaks as little against its origin in the pure, exclusive creativity of individual life as does the origin of physical offspring from no other potency than that of the parent. And just as the creation of this autonomous being, independent of its creator, is immanent to physiological life and, in fact, characterizes life as

such, so too is the creation of an independently meaningful content immanent to life at the level of the mind [*geist*].[38]

Here Simmel explicitly connects the worlds of biology and culture through the concept of creativity and the example of reproduction. Simmel's pan-vitalist vision depicts life producing more of itself even at the basic, vegetative level of physical reproduction and physiological self-maintenance. But while biological reproduction is a process that creates a wholly autonomous organism, at the level of mind life generates "more-than-life" in the form of the autonomous and valid content (*Gehalt*) that constitutes shared ideal worlds.[39] In the terminology of contemporary philosophy of mind, Simmel would be considered an emergentist with respect to the relation of value to the physical world. Rather than being debunked by Darwinian evolution, Simmel saw his conception of value as validated by it. Darwin presented nature as an endlessly creative process that generated novel forms of life that achieved temporary stability and fixity in the face of the dynamic flux of evolution. This same creative process of form and fixity was the key to understanding the place of value in nature.

In the final chapter of his *Lebensanschauung*, Simmel defines ideal worlds through the terms *meaning* (*Sinn*) and *form* (*Form*):

> Life at the level of mind [*Geist*] produces objective structures as its immediate manifestations, in which it expresses itself and which once again, as life's containers and forms, receive its further currents—as their ideal and historical fixity, boundedness and rigidity sooner or later come into opposition and antagonism with the ever-variable, boundary-dissolving, continuous life. Life is perpetually producing such things upon which it breaks itself, by which it is violated, things that are necessarily its appropriate form but which already through this, by being form, in the deepest sense conflict with the dynamic of life, with life's incapability of actually staying put. . . . This relation is the truly fundamental opposition between the principle of life and the principle of form, which, because life *can* only present itself in forms, is expressed in each particular case as the struggle of the form just brought forth by life against the forms that life has previously produced as its figures, its language, its determinate quality.[40]

This passage captures the Darwinian resonance of Simmel's conception of the creative dynamics of life.[41] The impossibility of stasis and the inevitability of conflict that results make such forms unstable, only temporarily or relatively valid. Life for Simmel is just this process of the creation of temporarily stable, determinate forms in which life expresses itself, followed by the destruction of these forms through the same creative principle. It is in this sense that Simmel calls his own metaphysical Darwinism, in Nietzschean fashion, a tragic view of life. The philosophical lesson of Darwinism and this view of life was the impossibility of breaking out of limitations, finitude, and the impermanence of both organic and ideal forms. His formula of life as self-transcendent thus declared that philosophy can no longer strive for the absolute, but must be content with only relative autonomy, and only relatively fixed, relatively valid values.

Religion As a Life-Process

Simmel developed his theory of religion in several essays from various periods in his career: "Contributions to the Epistemology of Religion" (1901); "Religion" (1906/1912); "A Contribution to the Sociology of Religion" (1905); "A Problem in the Philosophy of Religion" (1905); "On the Personality of God" (1911); and "The Problem of the Religious Situation" (1911). Although it is not possible here to do justice to the details and subtleties of these individual essays, the general structure of Simmel's approach to religion and its relation to the wider philosophical themes already mentioned can be found in each of them. In general, Simmel explains religion through the same creative, dynamic process of life that he later came to see as the source of the autonomous ideal worlds that constituted distinct arenas of culture, each answering to its own foundational criteria of validity—law, science, ethics, art, and the economy. Like economic values, religions involved a process of objectification by which psychological states of desire, emotion, and "feelings of value" (*Wertgefühle*) came to constitute ideal worlds. Simmel's theory of religion is finally genealogical—not aimed at rational justification or critique. However, because his genealogy is aimed at securing the notion that religions are a distinct and autonomous arena of cultural value, constituted by their own internal norms of justification and critique, he maintained that one could not engage in a critique of religion without doing so on the playing field of its own norms. And yet, just as economic values were seen to have their source in given subjective desires and needs that were not yet economic, so too religious ideals were seen to have their origin in subjective states and social relations that were not yet specifically *religious* in character.

Like Nietzsche and Overbeck, Simmel approached the unique nature of religion by identifying its specific difference from purely theoretical knowledge. He writes of the difference between "faith" (*Glauben*) and "knowledge" (*Wissen*):

> Belief [*Glaube*] in the intellectual sense stands in a series with knowledge [*Wissen*] as a merely lower stage of the same; it is a "holding to be true" [*Fürwahrhalten*] on the basis of reasons that are only inferior in terms of quantitative strength to those reasons on the basis of which we claim to *know*

something. In this respect metaphysical or epistemological investigations can lead us to hold the existence of God to be a plausible or, under certain conditions, necessary hypothesis. Then we would believe in God in the same way as we believe in the existence of light [*Lichtäther*] or in the atomic structure of matter. But we intuitively feel that when a religious person says, "I believe in God"—something else is meant than a certain "holding to be true" of God's existence.[42]

For Nietzsche and Overbeck, religious beliefs did not assert value-neutral facts but rather facts about value—indeed facts about the value of life itself. Similarly, Simmel argues that religions aimed not simply at knowledge that took the form of a neutral "holding something to be true," but at a picture of reality that made a practical difference. Simmel writes further, "Whether one believes in Jehova, in the Christian God, in Ormzud and Ahriman, in Vitzioutzli—that is not only with regard to content, but also functionally different; it proclaims a different kind of *Being* of the people."[43] For Simmel, science was a normatively constituted sphere of human activity that was characterized precisely by the independence of its purely theoretical aims from other arenas of human action. In this sphere, Simmel suggests, scientific knowledge of the world can change without this change having any necessary practical implication for other areas of culture. However, different ways of conceiving reality *religiously*—even the negation of certain ways of conceiving the religious character of reality—necessitate a different set of values, a wholly different way of life in the world. The problem of value thus afforded Simmel a novel perspective on the theory of religion, as it did for Nietzsche and Overbeck, because religious conceptions of reality answered to the question of the meaning of reality for *us*. This was a distinctively existential question that could only appear nonsensical from the purely theoretical and value-neutral perspective of science. While both science and religion were ideal worlds that were constituted by values, the epistemic values orienting science were of fundamentally different sort than the guiding extra-scientific and existential value that established the internal normative standards of religious life.

Just as Simmel characterized money as an ideal world of *instrumental* rather than *intrinsic* value, so he characterized religions according to the unique kinds of values they pursued. These were values of purity and holiness; reconciliation and social harmony between individuals and groups; cosmic justice in the face of the reality of evil; unity over the multiplicity of experience; and a longing for an absolute object that was to finally satisfy desire. Simmel argued that religious values expressed states of fulfillment and frustration, satisfaction and elevation, and, at the other extreme, even boredom that were infused with the significance of the final value of life.[44] These states were the affective forms through which diverse representational content of experience took on a religious meaning. Simmel argued that "religiosity"—perhaps what we might today call "spirituality"—was a distinct "life-process" (*Lebensprozess*) underlying the values that were embedded in the religious dimensions of lived experience and according to which the world came to take on the contrast of the sacred and the profane.

Simmel held that the matter that religiosity comes to shape into its unique ideal world actually resides in the immanent world of experience, not in a transcendent realm. But religiosity was a natural tendency of human subjectivity to shape features of the experienced world in relation to the idea of transcendence, of a higher reality beyond experience. Simmel writes,

> Religiosity, as the innermost condition of life, as the incomparable functional mode of certain realities, first robs, so to speak, a substance for itself in the course of its tour through the diverse content of the world and thereby places itself over against itself, the world of religion over against the religious subject. It must first color the content of the world through its manner of experiencing, leaving other ways of realizing this content behind, in order to then to build the countless worlds of belief, of gods, of salvation [*Heilstatsachen*] out of its now accumulated religious value.[45]

In the experience of nature, of fate, and of social relations, religiosity finds the material that it then fashions into an ideal world of vital concern and ultimate value. Affects of respect, love, fear, and admiration, uncertainties of fate and complexities of relationships with other human beings take on a religious character by becoming charged with the ultimate significance of life's own deeper striving.

The goal of Simmel's theory—like Nietzsche's before him—was to offer a purely immanent genealogy of the human orientation to otherworldly and transcendent realities. Simmel writes,

> [A]ll of these [needs] nourish transcendental representations: the hunger of the human being is their sustenance. The believer in the most pure religious sense does not pay attention to its theoretical possibility or impossibility, but instead feels exclusively that his desire finds its outlet and fulfillment in his belief.... What is essential is that transcendent realities are thought and sensed at all, and their truth is only the immediate and completed movement that leads to them—roughly as a strong, subjective sensation compels us to believe in the existence of an object that agrees with it, even when we must logically doubt this [existence].[46]

The metaphysical hunger of human beings is the vital foundation of ideals that the individual encounters as given autonomously, objectively, and transcendentally. Simmel argues here in agreement with Nietzsche and Overbeck that religious truth is in part constituted by the intensity of experience in the valuing subject. The source of *validity* for religious values—the internal norm constitutive of justification and criticism—is the felt intensity of life that they are connected with; not an objective order of things represented disinterestedly, but the inner intensity of the subjective act of valuing itself. In one sense, Simmel utterly psychologized the validity of religious ideals and values; in another sense, he grounded them in the positive, given reality of the internal, appetitive, and erotic life of subjectivity. And this appetitive psychic life manifested the creative drive behind the organic world as a whole.

As his essays on religion show, Simmel used the practice of science (*Wissen*) as the foil through which to approach the question of what religion is. Simmel defined science too according to the constitutive value at which it aims, and this was the value of theoretical knowledge *for its own sake*. Like Nietzsche and Overbeck before him, Simmel puzzled over how to understand the emergence of such a value—which appeared to serve no "vital-teleological" purpose—from nature. He wrote:

> The character of all science as such seems to me to consist in the fact that certain mental [*geistig*] forms exist ideally (causality, inductive and deductive reasoning, systematic order, criteria for establishing facts, etc.) which the given contents of the world, through placement into them, must satisfy. Expressed in psychological terms: at first humans know in order to live, but then there are humans who live in order to know. Which *content* is selected to show oneself as one who fulfills that demand is truly accidental and depends upon historical-psychological constellations: for science all contents are in principle equivalent, because with science, in its most characteristic contrast to vital-teleological knowing, the individual object as such is a matter of indifference.[47]

It is precisely the *aim* and *ideal* of scientific knowledge, and the distinct manner of representing an object that fits this aim, that characterized it as an ideal world. In Kantian idealist fashion, this aim, for Simmel, consists in fitting the contents of experience into *a priori* and ideal forms of intelligibility. Science distinguishes itself from all other forms of human cultural life (i.e., art, religion, law) because in it truth is sought for the sake of truth, generality is privileged over the uniqueness and diversity of experiences and objects, and "the individual object is a matter of indifference."[48] Although Simmel contends that the norms of scientific cognition themselves have their origin in life, the defining feature of science is its unique *interest* in objects only insofar as they comply with these epistemic norms and aims. In the order of evolution, cognition is first in the service of life's vital drives, but the very evidence of life's self-transcendence and the autonomy of ideal worlds is the fact that life then comes to serve as a means to these autonomous ends, valuable *for their own sake*.

Simmel applied this framework equally to all of the ideal worlds of culture whose contents were infused with "spiritual" (*geistige*) activity. Their mark was precisely that their constitutive aims could no longer be explained according to the "vital-teleological" needs of the biological world out of which they emerged. Instead, Simmel referred to all of these cultural domains as "series of purposes" (*Zweckreihen*) that are characterized by a new form of teleology and purpose that is independent from that of life.[49] These cultural spheres have achieved relative autonomy from the merely biological teleology—they have become "more than life."[50] Art, science, religion, the economy, and law are evidence of the autonomy of *Geist* from *Natur*, of value from life, precisely because the ends that they seek to realize are no longer means to "vital-teleological" ends that are biologically given; instead, they subordinate biological life to their own unique and constitutive purposes.

This conception of science and religion has implications for understanding the nature of their conflict. The scientific value of theoretical truth *for its own sake* could come into conflict with other spheres, but this conflict was not merely on the first-order level of conflicting claims about the natural order of cause and effect. Most fundamentally, it was also on the second-order level of competing *criteria* of normative validity. Each arena generated a picture of reality in conformity with its own guiding normative criteria. As Simmel writes, "In practice, truth is sought for the sake of life; in religion, for the sake of God or holiness; in art, for the sake of aesthetic values. If within these last two teleologies other notions are more useful than the *theoretically* true ones, then these others are sought instead of the latter."[51] The true conflict between these spheres occurs at the level of purposes and aims, when any one of them legislates its own guiding value as the highest or final aim to which the rest must be subordinate. Pursuing any one of these values exclusively, Simmel thought, requires subordinating the orienting values of other spheres. The true conflict that these diverging spheres presented thus manifested itself in the internal life of individuals, who were forced by the necessity of leading a life to choose between conflicting ideal worlds whose values were incommensurable and could not be simultaneously realized. Simmel thus recast the conflict between science and religion from a conflict between contrary theoretical claims about the causal order of nature to a conflict between orienting values. Science, dedicated to truth *for its own sake*, and religion, dedicated to the final unity of being and value, demanded wholly different ways of life.

Conclusion

Simmel's theory of value in *The Philosophy of Money* established the theoretical foundations for a broader thesis that the decisive factor in shaping the values of modern culture was the ever-more-dominating money economy. He argued that the abstractness of money as a measure and criterion of value fueled a process of increasing social and economic abstraction and rationalization that in turn led to the dominance of instrumental over intrinsic values. This ushered in a new "revaluation" of values in modern culture, in which intellectualism and a new conception of rationality—one that viewed reason purely in terms of instrumental value—dominated political and ethical life. However, Simmel found that the money economy also had a social effect, thanks to the increasing opportunity to accumulate private wealth. Private wealth allowed individuals to sever their ties of material dependency on specific social groups, families, and professional guilds. Individuals could thereby move through an increasing variety of social spheres and even opt out of them altogether. Simmel argued that these economic conditions actually led to an increase in a negative form of individual freedom—namely, freedom to sever ties with social groups and to withdraw from participation in particular spheres of society.[52] But while this new independence increased negative freedom, the abstractness and instrumentality of social relations in the money economy resulted in increasingly impersonal and instrumental social relations.

The increased individual freedom realized through the money economy led to another form of conflict. Coupled with the functional differentiation of cultural spheres in modern culture (art, science, religion, economy, law), the freedom to move about in many spheres led to individuals becoming subject to socialization in increasingly fragmented and incommensurable ways. The incommensurable plurality of cultural values in economic, political, religious, scientific and aesthetic domains—as well as the plural values within each of these spheres—pulled individuals in opposing directions. This created new forms of cognitive dissonance and crisis in the internal lives of individuals as they internalized the disparate values without having a way to order them under a single ultimate value.[53] The increase in negative freedom to move through different value-spheres thus also had an effect on inner life as individuals internalized the conflicting values of different ideal worlds.

Simmel finally diagnosed this new cultural situation through an appeal to Nietzsche's notion of tragic philosophy. The "tragedy of culture," as he called it, was the inevitable alienation of cultural forms from the very needs, desires, and impulses that generated them and that were to be fulfilled in them in the first place. This tragic dynamic was itself an inevitable result of the creative cycle that drove life to take on new forms.[54] Simmel wrote, "By a tragic fate, we surely mean this: that the destructive forces directed against some being spring forth from the deepest layers of this very being. It is the very concept of all culture that the mind generates something autonomous and objective, through which the development of the subject from itself back to itself receives its path; but the integrating, cultural element is predetermined to follow its own unique development, which continues to use up the powers of its subjects and to pull them into its path without thereby raising them to their own apex."[55]

In short, the individual was caught in the current of conflicting value spheres that took on their own momentum, autonomous from the sources in subjective life that generated them. In this process, individuals themselves became mere means to the now autonomously evolving social forms that could no longer promise to fulfill them by "raising them to their own apex."

For Simmel, the dynamic process of life out of which values and ideals were generated suggested that even human nature, with its diverse collection of needs, desires, and spiritual hungers, was thoroughly in flux. The concept of human nature had no content, and there was no basic, natural value—even of life itself—that could be drawn from the nature of the organic world. Yet, Simmel, like Nietzsche, drew a final lesson about values from this situation. He argued that, rather than eroding all notions of value, this new conception of life could be the soil for a new kind of *ethos*. The ethical ideal appropriate to this image of the place of humanity within the creativity of nature could not be Kant's universalizing "categorical imperative"; rather, it had to be a unique "individual law."[56] This ethical goal of Simmel's Darwinian Life-philosophy is present as early as his *Einleitung in die Moralwissenschaft* and persists throughout his work. It became the ethical principle of his final work. The last chapter of his sweeping metaphysics of creative life was titled "Das individuelle Gesetz" (The individual law). The "individual law," as opposed to the categorical imperative, urged each individual to pursue her own creative solution to the tragedy of culture, to the challenge of pursuing conflicting values amidst the diverging spheres of modern society.

In the end, Simmel too, came to see the Darwinian conception of evolving life not as a nihilistic or debunking form of naturalism, but as a resource for a new *Weltanschauung* in its own right. This new worldview fused the tragic metaphysics of life, which stood behind the creativity of both nature and culture, with an ideal of individuality and the free, ungrounded, creative synthesis of values that every individual must perform in her unique appropriation of the values constituting the ideal worlds that she faced. Simmel too succumbed to a religious temperament by championing creativity and individuality as ways of life that were most in harmony with the creative pulse of nature itself. The ideal of the individual law aimed to offer individuals a new and more realistic sense of what forms of human flourishing were still available in the dynamic modern world. In creativity, each individual became a microcosm of the dynamic drive of life for "more life and more-than-life" that pulsed through both nature and culture.

Simmel wrote that the aim of his philosophy was to "place back into life everything that had been established outside of life," and by doing so to offer a way of understanding what meaning nature had for the human pursuit of values.[57] This formula serves to capture the basic philosophical aims of Life-philosophy in general that have been presented in these three chapters. Simmel was the culmination of the Nietzschean attempt to marry philosophy and Darwin, to show what meaning the evolutionary view of nature had for the diverse values and traditions of European culture. Moreover, Simmel more than Nietzsche or Overbeck self-consciously aimed to defend the philosophical import of evolutionary biology against forceful criticisms from the neo-Kantians that will be encountered in the next chapter. Each of the thinkers met with thus far has attempted to show how the evolutionary picture of a dynamically creative and destructive nature could recast the relationship between the values found in scientific investigation and in religious life. The vital drives that humans share with the rest of nature—a biological *eros* that permeates all living things—offered a theory of the source of the normative power of ideals that captured human agency. Finally, life became the pillar of a new worldview that challenged both the lingering rationalism of the classical German philosophical tradition and the scientism of the academic *Wissenschaften*. The views of Nietzsche, Overbeck, and Simmel regarding *Glauben* and *Wissen* suggested that religions conflicted with science in the same way that science conflicted with life in general: namely, at the level of the guiding values that these spheres pursued. Insofar as religions were orientations toward the value of life as a whole, and not toward the value of truth in and only for itself, they were extra-scientific. But in Simmel's view, they were nonetheless vital.

4

Heinrich Rickert: The Autonomy of Agency and the Science of Life

The philosophical aim of all three Nietzschean Life-philosophers was nicely summed up by Georg Simmel's wish "to place back into life itself everything that had been established outside of life."[1] On the other side of the debate about the place of values in nature in this context stood the Southwest "Baden" school of neo-Kantianism that began with Wilhelm Windelband, a student of the late-Idealist philosopher, Hermann Lotze. This lineage of neo-Kantian thought culminated in the work of one the more neglected thinkers of early twentieth century, Windelband's student and protégé Heinrich Rickert (1863–1936). A professor in Freiburg starting in 1894 and then Heidelberg in 1915 until his retirement in 1932, Rickert was a towering figure in fin-de-siècle academic philosophy who established his reputation through rigorous attacks on positivism, naturalism, historicism, and the alliance of philosophy with the sciences. He understood his philosophical mission to be the defense of a broadly Kantian, transcendental idealist conception of philosophy, for which the very concepts of philosophy and science depended on a proper understanding of the nature of value and its role in various arenas of human activity, including science and religion.

Rickert's significance can be indicated by the list of diverse figures who were influenced by him and who have been influential in their respective fields. Max Weber and Ernst Troeltsch were deeply indebted to Rickert's methodological writings on concept formation in the human sciences, and Troeltsch would even claim Rickert as an inspirational interlocutor in his philosophy of religion.[2] Georg Simmel, Wilhelm Dilthey, and Karl Jaspers all took Rickert's transcendental theory of value to be a position against which they defined their own, as did some of Rickert's more famous students: Emil Lask, Martin Heidegger, and Walter Benjamin. Edmund Husserl and, on the other side of the Atlantic, William James developed their enormously influential schools of phenomenology and pragmatism in critical conversation with Rickert's brand of Kantian idealism. In addition to this influence, of course, Rickert's engagement with Nietzschean Life-philosophy and philosophical Darwinism, which is the main subject of this chapter, also influenced philosophical assessments of naturalism throughout this period. While Rickert's many targets of critique meant that he too had many critics, his critics recognized that he had clearly identified the crucial issues, and had acutely defined the terms of debate.

The Baden neo-Kantian tradition that culminated in Rickert is typically contrasted with the Marburg school. Starting in the 1870s, these two schools sought to defend the insights of Kant and reclaim Kant's legacy in the context of the growing influence of the natural and human sciences. The Marburgers—Hermann Cohen, Paul Natorp, and Ernst Cassirer—are usually represented as philosophers of science and mathematics who were less interested in the wider cultural implications of philosophy or its place in cultural history. Their concern, so the traditional narrative goes, lay in methodological precision and rigor, in reminding contemporary scientists of the *a priori* logical forms and normative content invoked in the sciences and mathematics that were required for distinguishing *valid* from *invalid* thought. This characterization simplifies things, as all of these Marburgers also addressed the social, political, economic, and even religious issues of their time. Indeed, all were interested in the extent to which philosophical issues of normativity and validity were at play in and could potentially influence cultural debate about values in general. Ernst Cassirer, in particular, would go on to develop a powerful neo-Kantian theory of *a priori* symbolic forms that could unify the natural and human sciences and also offer insight into the cultural significance of religions.[3]

For these reasons, the division between the two schools is difficult to capture neatly. On the one hand, it is helpful to think of it as a product of genealogy. Herman Cohen studied under another late Idealist, Friedrich Trendelenburg, in Berlin and moved to a professorship in Marburg, where he taught Paul Natorp, who taught Cassirer. Windelband studied under Hermann Lotze and taught Rickert in Strasbourg, and Rickert then taught in Freiburg and Heidelberg. On the other hand, the difference between these schools is reflected in different sorts of emphasis in a common set of problems. Both schools shared the aim of providing a theory of *validity* (*Geltung*), and this term, first introduced by Lotze, captures the central aim of neo-Kantian philosophy perhaps better than any other. The problem of validity was the problem of how to understand the irreducibly normative and prescriptive content of principles that were embedded in all arenas of thought and action—including logic, science, ethics, religion, and even art. The Southwest, "Baden" tradition emphasized Kant's distinction between capacities or faculties—sensibility, understanding, reason, and desire—that gave rise to distinct domains in which unique normative principles operated. They stressed the significance of Kant's conception of regulative ideas of reason that guide thought and action toward a completion of each of these capacities, and this notion played a crucial role in Rickert's critique of philosophical Darwinism. This different emphasis on the metaphysical and regulative ideas of reason over the constitutive principles of natural science and mathematics also gave the philosophy of religion special standing in Rickert's thought. There is one more term that can be used to differentiate these schools: the concept of "value" (*Wert*). Value was another definitive, central concept of the brand of neo-Kantianism developed by Windelband and Rickert, which came to be known simply as *Wertphilosophie* (the philosophy of value).

Rickert began writing in the 1890s at the height of the famous *Methodenstreit* over the proper demarcation criteria between the human sciences (*Geisteswissenchaften*) and the natural sciences (*Naturwissenschaften*). Rickert established a unique position in this debate by arguing that the natural and human sciences are to be distinguished not by the different types of "stuff" in the world that they study, but by their different *aims* and the methods that were appropriate to realizing these unique aims. At the end of Rickert's career, his essays focused on metaphysics, religion, and the problem of acquiring a philosophical worldview (*Weltanschauung*) that might orient practical life in an age dominated by the theoretical pursuits of the sciences. These late essays aimed to offer a neo-Kantian rejoinder to the growing popularity of Martin Heidegger, who was a former doctoral student of Rickert and whose program for phenomenology in the 1927 work *Being and Time* had exploded in the interim. Rickert's career in academic German philosophy thus spanned a vibrant, creative, and dynamic period whose ripples are still felt. But within this period of great change, Rickert always understood himself as trying to hold and defend a strict Kantian line that stayed true to the lasting insights of Kant's critical philosophy against those who had failed to digest its radical message or who wished to go beyond its limits.

Rickert's critique of biological thinking, Darwinism, and Life-philosophy was part and parcel of his defense of Kantianism, and it signified an important turning point between his early works on methodology, philosophy of science, and epistemology, and later works on culture, religion, and metaphysics. In this chapter, I introduce elements of Rickert's theory of value first through an analysis of his critique of Life-philosophy in his 1920 monograph, *Die Philosophie des Lebens: Darstellung und Kritik der Philosophischen Modeströmungen unserer Zeit* (The philosophy of life: a presentation and critique of the philosophical fashions of our time). Next, I place this criticism within the context of Rickert's wider project of a non-naturalist, Kant-inspired understanding of philosophy as a theory of *valid* values. Finally, I turn to Rickert's conception of the relationship between philosophy, science, and the existential problem of securing what he called a "worldview." My overall aim is to examine Rickert's critique of the way the Nietzscheans had appropriated biological discourse to understand the place of value in nature in light of Rickert's own unique mediation between science and religion.

Against Fashionable Philosophy

Rickert's critique of Life-philosophy is outlined in detail in his fascinating 1920 work of both cultural polemic and trenchant philosophical argumentation. The book was published two years after the death of his personal friend, Georg Simmel, who, in his estimation, had presented the most consistent and compelling version of philosophical Darwinism available. It is a piece of writing steeped in the late-Idealist, post-Hegelian, German philosophical tradition that targeted the chaotic mix of nineteenth-century "-isms:" existentialism, vitalism, naturalism, intuitionism, irrationalism, and Romantic rejections of rationalism and systematic philosophy. Besides Nietzsche and Simmel, Rickert cites a long list of familiar thinkers as belonging to the "fashionable trend" of Life-philosophy and exhibiting what he took to be its central errors: theorist of the human sciences Wilhelm Dilthey (1833–1911), French vitalist Henri Bergson (1859–1941), phenomenologist Max Scheler (1874–1928), and a host of Darwinians including Darwin himself and Herbert Spencer.[4] Rickert saw his settling of scores with Life-philosophy as providing the definitive nail in the coffin for—in his view—all of the confusions surrounding the mixture of the natural sciences, especially the life-sciences, and philosophy that had been steadily spreading throughout the nineteenth century. The Nietzschean appropriation of Darwin was just one example of a wider infection of philosophy by a worrisome brand of scientism: the encroachment of the sciences on the problems of philosophy, which the sciences were in no position to solve. Indeed, Rickert's critique also offers one of the first intellectual genealogies of Life-philosophy itself, tracing it back to the counter-Enlightenment reaction against Kant—especially to romanticism, Goethe, Schopenhauer, and the "inventor" of the problem of nihilism, Friedrich Heinrich Jacobi.[5]

While Rickert's *The Philosophy of Life* is a above all a work of argumentation written by a philosopher who stressed clarity and systematic rigor above all else in philosophical thinking, it also contains a tone of polemical, passionate critique that resembles the style of one its central targets: Nietzsche. Indeed, one can read Rickert's polemic against Life-philosophy as an early precursor to commonly stated, and often equally polemical, contrasts between "Continental" and "Analytic" philosophy. For Rickert, the Life-philosophers sacrificed clarity and rigor for the sake of the impression of depth and for emotional, affective appeal. Of course, for the Life-philosophers, Rickert's sober, academic style was dry and lacked the *vital* concern and urgency out of which philosophy itself originally emerged and to which it had to respond in order to be existentially relevant. The divergence between these approaches is scathingly captured by Rickert's epithet for the Life-philosophers: They were *Wertpropheten* (prophets of value) rather than sober, rigorous, and "scientific" *Wertphilosophen* (philosophers of value).[6]

Rickert polemically discredited Life-philosophy as "pop" philosophy, in all of the pejorative meanings of the word. It was a philosophical fad and flavor of the month, whose stark inconsistencies and logical inadequacies would soon be apparent to all once the baffling spell cast by the youthful, exuberant passion that animated it had worn off. Rickert's critique was, of course, intended to aid in the process of breaking the spell and bringing all of the ambiguities and fallacies that he saw hidden under the catch-all term *life* to the light of day. In his preface, he even mockingly wrote of Nietzsche that "if I am allowed nonetheless at this point to make an assumption about the future, then it would be to the effect that we are at the end of the philosophy of mere life. I would be glad if this small book could be regarded as an indication of such a 'twilight of the idols.'"[7] This clear reference to Nietzsche as a prophet of life is echoed as Rickert soon after writes, "Maybe we must first to go through Hegelianism as well, before we resolve ourselves again to practice self-reliant [*selbstständig*] philosophy, and in any case there is more to learn about the timeless problems from Hegel than there is from Zarathustra."[8] Rickert's work was supposed to move philosophy away from youthful enthusiasm and fashionable appeal to sober and careful reasoning about to the "timeless problems" that were its only justifiable interest.

At first, Rickert scorns what he sees as an empty championing of the living over the dead in all areas of cultural life, from religion, art, and literature to science and philosophy themselves. The enthusiasts of life, he argues, use it as the universal norm and value, calling everything that they simply feel affinity toward living and alive. This all too easily fed into a superficial presentism that masked purely arbitrary declarations of "boo" and "hoorah," even toward philosophy itself. The danger of such empty rhetoric of life, Rickert claimed, was that it obfuscated the real aesthetic, scientific, ethical, and religious values that lay behind such empty evaluative terms as *living* and *dead*. This point was especially clear for Rickert in the case of science, for he thought it was patent nonsense to champion a so-called living physics or mathematics at the expense of a dead one. The concept of life could never be useful for clarifying the norms and values actually at play in the various sciences—for example, the norms that one appeals to in favoring one physical theory over another. For Rickert, the only normative criterion that it made sense to invoke in scientific inquiry was finally that of *truth*, which he took to consist in the match between theory and experience; the consideration of whether or not a theory encouraged or enhanced life was nonsense. Rickert polemically writes in response to the "prophets of life" that "it is not unusual that [the concept of life] represents an empty phrase and serves as a cloak [*Deckmantel*] for thoughtlessness."[9]

Life-Philosophy and Life-Science

Despite this dismissive prolegomenon, Rickert's book quickly moves from polemic to the "timeless" problems that he insisted should be the main concerns of philosophy. In light of the last three chapters, it is obvious that these dismissive remarks could only be directed at "straw-men," since the Life-philosophers too clearly recognized the status of the value of truth in the sciences. The Nietzscheans canvassed in the previous chapters saw science as a philosophical problem precisely because its central value of truth *for its own sake*, at first glance, did not appear to have any precedents in the activities of living things more generally. Life was not to supplant more concrete cultural values or the norms that constituted scientific inquiry but to identify their sources in nature and to expose the nonrational factors that gave rise to them and undergirded, finally, their normative appeal. Their basic aim was meta-theoretical, and their interest in biology sprang from the effort to relate human norms and values to the teleological character of evolving living things. In this sense, their inquiry too was transcendental in that it turned to biological life-processes in order to investigate the conditions of the possibility of value and valuing in nature at all.

As we see in the quotation above, Rickert criticized Life-philosophy for failing to achieve the status of truly "self-reliant" philosophy. This criticism is key, and it reveals some of Rickert's own foundational assumptions. A "self-reliant" philosophy was one that could redeem the normative *validity* of its own presuppositions. Because Life-philosophy looked to biological science and even to nature itself for the sources of normativity, it could only take the authority of norms and values to be something that is given and dictated by nature, independently of rational thinking. The failure of Life-philosophy in this regard, as he saw it, was evident in the bundle of doctrines it jumbled together. These included intuitionism in the theory of knowledge; biologism as a metaphysical, meta-ethical, and normative ethical doctrine; the rejection of systematic philosophy; and, finally, a naturalistic and genealogical rather than a genuinely transcendental and rational understanding of value. Rickert argued that this incoherent bundle of doctrines could only result in skepticism about the very possibility of "self-reliant" philosophy, and this skepticism could only reinforce a nihilistic rejection of the genuine validity of the norms and values that constituted the arenas of cultural life and scientific inquiry. Rather than vindicating the new ideals championed by our Life-philosophers—the "affirmation of life" (Nietzsche), "freedom" (Overbeck), and the "individual law" (Simmel)—the very principle of life used to support them also undermined them. It even undermined the validity of its own representation of life.

To understand these bold accusations, it will help to say a few words about Rickert's use of the term *biologism* (*Biologismus*).[10] Rickert uses this term to describe any theory that takes central concepts of biology, the science of life, as the basis for either theoretical philosophy, which included metaphysics and epistemology, or practical philosophy, which included ethics and political theory. He saw this as the philosophical fashion that he aimed to challenge, but he distinguished an older biologism from a newer one. To the old variety belonged "traditional" Darwinists in the mostly English context, a group in which Rickert included Darwin, Herbert Spencer, and the political economist Thomas Malthus, whose views on population dynamics and carrying capacity were so influential for Darwin's formulation of the theory of evolution by natural selection.[11] The principle feature of this "older" version of biologism was the widespread use of principles to explain physical processes, biological traits, and evolutionary change—such as natural selection, adaptation, self-preservation, organic development, or even the law of the conservation of energy taken from physics—to explain the origins of social, cultural, and even scientific values. He found these to be crude and invalid transferals of biological, and even physical, concepts to the realm of human agency and action, but even more problematic was the use of these concepts to assert and even assess normative validity claims.

Rickert takes on the "older" biologism first by pointing to the fact that these principles were used during his time to support values and policies from across the political spectrum from Left to Right, from socialist to individualist, anarchist to aristocratic, democratic to despotic. The application of these biological concepts to questions of normative ethics and political thought allowed such diverse appropriations, Rickert thought, because they were generally empty. The principle of natural selection or the law of conservation of energy could yield no concrete guidance if taken as principles of practical reasoning that are to tell us how to act and how to shape society. As to the application of such principles to questions about the origin and nature of religion, Rickert objects that these led to crude explanations of religious beliefs and practices as strategies for enhancing reproductive success rather than as attempts to achieve ethical wisdom or metaphysical knowledge.[12] The wide range of ideologies that marshaled Darwinian concepts for normative ends only indicated for Rickert the confusion surrounding the distinction between the positive sciences and explicit inquiry into valid normative principles.

In addition to the "old" Darwinism of the primarily English context, Rickert pointed out that there was a "new" Darwinism that rejected what was perceived as an overly *mechanistic* understanding of nature. The "new" Darwinism instead pointed to the biological world as requiring an altogether different set of principles. This approach was typified in "vitalist" biology and especially in the works of Continental Darwinists like the Nietzscheans, who saw living things and the evolutionary process as fundamentally directed, not necessarily at human life or consciousness, but at abstract and spiritualized goals, such as Simmel's "more life and more-than-life." Living things were not driven only by material or even evolutionary interests related to self-preservation, the battle for resources, and reproduction. They were also driven by an appetite and desire that permeated the creativity of the biological world—Nietzsche's biologized *eros* of life desiring itself, Overbeck's vital forces, Simmel's notion of "self-transcendence," or Henri Bergson's notion of *élan vital*. Rickert recognized that this "new," Continental Darwinism yielded different values from those of the mechanistic Darwinists. Ethical individualism, organicist theories of the state, and even, as we saw especially in Nietzsche, a "revaluation" of the value of science and morality themselves could be seen to harmonize with this view of life.[13]

Rickert linked this "new" biologism to an epistemological doctrine that he called intuitionism. The antimechanistic, antimaterialist Darwinists, Rickert claimed, regarded *intuition*—and not abstract, logical, and conceptual thought—as the source of the apprehension of organisms as what they genuinely were: teleologically organized wholes. Further, only through intuition could the whole of nature be apprehended as a unified, teleologically organized whole. While the analytic methods of the sciences and the abstract concepts of the intellect "violently" carved up nature into distinct, disconnected, and mechanical parts, intuition was the faculty by which nature as a whole in its organic unity could be discerned.[14]

Life in the sense of vitalist Darwinism promised to show how "spiritual" (*geistig*) values no longer tied to material biological needs could emerge from nature, whereas the "older" version only offered crude reductionism. It was also to achieve the intellectual goal of providing a *unified* theory of *Sein* and *Sollen*, "being" and "validity," by seeing everything—normative values included—as the product of an immanent, creative principle manifested in organic forms of life. For Rickert, Life-philosophy promised the ultimate philosophical *desideratum* of a monistic unification of mind and nature (*Geist* und *Natur*), for which a set of orienting values was written into the common foundation of both. For this reason, the life sciences were a route to satisfying not just the intellectual desire for theoretical knowledge of the causal order and constitution of life, but the very existential desires that the Life-philosophers saw embodied in religions. The life sciences could reveal a genuine path to human flourishing and a way of life in harmony with the creative powers of nature.[15]

Rickert's critique of these biologistic ambitions, in both their "old" and "new" forms, had two essential parts: the first was the criticism of biologism as a doctrine of *being* (*Sein*), and the second was a criticism of biologism as a doctrine of *value* (*Sollen*). The two angles targeted the adequacy of the concept of life for achieving the cardinal aims of theoretical and practical philosophy respectively: namely, to grasp the world conceptually as a coherent whole and to answer the question of what norms and values are valid. Rickert took the theoretical and practical aims served by the concept of life to be independent, but, like Simmel, also drew attention to the way in which the Life-philosophers aimed to fuse them. Rickert argued that the two domains of *Sein* and *Sollen* were fundamentally logically distinct, but he grounded this logical distinction in the peculiar nature of the human capacity for self-legislative agency. Even Simmel's notion of life's emergent self-transcendence into a realm of valid values could not overcome the fundamental difference between judgments about sources in the sense of *causal* origins and judgments of sources in terms of *justification* and *reasons* that normatively regulate thought and action.

Rickert first addresses the inadequacy of biology for yielding a comprehensive doctrine of *Sein*. His point here is rather simple: life cannot become a concept for reality as such because not everything in the world is living.[16] The principles that describe and explain the biological world, such as evolution by natural selection, or the concept of organism, for example, do not apply to parts of the world studied by other sciences. The pan-vitalist extension of biological principles to the world as a whole was an example of what Rickert insightfully identified as the fallacy of "specialist universalism" (*spezialistischen Universalismus*).[17] This was the widespread temptation to generalize principles that apply to *part* of reality to reality as a *whole*. Biolog*ism*, in contrast to the science of biology, takes the *part* of the world studied by biology—namely, the arena of life—to be the *whole* world (*das Weltganze*). Rickert writes, "The part of the world that [biology] limits itself to will never be anything other than a part. Within the discipline of biology, there is no doubt about this. Biolog*ism* as philosophy, on the other hand, wants to make the part into the whole."[18] The principles of biology as the science of life do not necessarily hold for the world studied by physics, chemistry, mathematics, logic, philosophy, or even theology, and thus they cannot satisfy one of the cardinal ideals of reason in distinction to the special sciences: to stitch together a picture of the world as it hangs together as a unified whole.

The notion of philosophy that Rickert deploys in this criticism is crucial for understanding his view of the inadequacy of biologism, and it is important that the very concept of philosophy be determined by an understanding of its unique *aim*. Philosophy, in particular the branch of philosophy called metaphysics, is "the universal science of the world as a whole," or in another passage, "universal knowledge of the real world."[19] Life-philosophy, like many other forms of reductionism in the natural and historical sciences, had overplayed its hand in attempting to create a unified picture of the world-whole through principles developed from the study of one unique part of the world. Rickert saw "specialist universalism" as a constant temptation toward unification that created a unique overlap between the ambitions of the sciences and the traditional ideals of philosophy. But Rickert was an early antireductionist and theorist of the disunity of science, who argued that the principles and concepts that sciences used and needed to conceptually grasp their parts of the world were often logically distinct and incompatible. There was no universal science or universal scientific method, only the Kantian regulative idea of a world-whole that pressed areas of specialization to seek greater unity. But the Kantian idea was that this world-whole was a project and ideal of reason. Thus, it was a specifically philosophical task to seek an understanding of how the results of distinct sciences could be built into a conception of nature as a whole.

Rickert's case against both "old" and "new" forms of biologism moved from the topic of metaphysics to the topic of epistemology. Neither a mechanistic biology, nor the vitalist version, nor any of the positive sciences, he argued, offered any account of how the central aim of science, objectively *valid* knowledge, was possible and achievable at all. Moreover, despite the fact that the sciences each presupposed the value of knowledge *for its own sake*, they could offer no account of how this value could be normatively valid. Rickert acknowledged that there had been attempts to develop a biologistic epistemology, and here he names American pragmatism as a theory of knowledge as *utility*, Simmel's conception of an evolutionary epistemology that fed into his late metaphysics of life, and prevalent Darwinian theories of knowledge as *adaptive success*.[20] None of these views, he suggests, redeems what science is—namely, the pursuit of knowledge *for its own sake*. The central epistemological questions of what knowledge is and whether it is possible, and the normative question of whether or not it is indeed valuable for its own sake, were not biological problems. Rickert argued that both the "old" and "new" Darwinian philosophies lacked an epistemology that could vindicate the validity of their own representations of life, the guiding aims of their own epistemic projects, and their implicit commitment to the intrinsic value of this knowledge.

Having rejected the "old" and "new" biologisms as viable theories of *Sein* and of knowledge, Rickert turned to the biologistic doctrine of *Sollen*. His critique unfolds in three stages. The first stage insists that biology, qua purely theoretical science, does not and *ought not* directly invoke value concepts in its explanations and descriptions of the world of *Sein*. Rickert insists that even central biological concepts like "adaptation," "development," "selection," "ascension," and "decline," which might appear to have a value component, have a value-neutral meaning and function in biological theory. In natural science, they can and *ought* to be formulated in a value-neutral way. This notion that there is an ought that governs the very notion of scientific representation emerges again throughout this chapter and is discussed further in the conclusion of this book. Life-philosophy, and this is especially pertinent to Nietzsche, took advantage of the double duty that the above terms can easily serve, thus obfuscating the distinction between their purely descriptive and regulative meanings.

The primary examples Rickert gives to illustrate the problem here are the master concepts of Nietzsche's cultural diagnosis: "health" and "sickness." These medical terms, which became rampant in nineteenth-century German thought and cultural criticism, have a purely descriptive meaning when applied to the functioning of a living system, so Rickert claimed.[21] But they can also function as concepts of value that regulate agency. Rickert writes, "From the perspective of biology, the concepts of ascending and declining life do not constitute an opposition of values, *and healthy and sick are no longer purely biological concepts insofar as one understands them under value or lack of value [Unwert]*. When a person is sick, bacteria live, and when the bacteria die, the person becomes healthy. *It is certainly a matter of the human will to take sides here and to set human health as a goal [Zweck]. . . . Such taking sides is foreign to the natural science of biology.*"[22]

For Rickert, it is because the medicalized master concepts of Nietzsche's evaluative project (e.g., "health" and "sickness") and the master concepts of Darwinian explanation (e.g., "selection" and "adaptation") can have both descriptive and evaluative meanings that they *seem* to overcome the divide between descriptive and normative claims. But this appearance is the root of their mistake, which consists in overriding different functions, practical and theoretical, that concepts can have. In other words, concepts can either describe mind-independent events in nature or they can regulate human agency. It is one thing to describe the causal organization of a system and the functional roles of its parts, another to "take sides" and aim at functioning as an end or claim that it is a valid value that *ought* to be nurtured, cared for, enhanced. The use of such multivalent and medicalized terms hides the two different logical roles they can play in rational agency, one pertaining to purely theoretical knowledge and the other to action and ethics.

Rickert's aim in sharply distinguishing these two senses is not only to protect the integrity of the descriptive empirical sciences, which he thinks have suffered from confusion surrounding the misuse of value concepts, but also to guard against the tendency of concepts in the sciences to slide surreptitiously into value-laden and ethical discourse. Concepts like "health" and "sickness" are paradigm examples of such slippage between logically distinct modes of thought because the values embedded in such terms appear self-evident (i.e., "health" and "life" are taken for granted as good things). This allows the switch that has taken place between the scientific and the agential applications of these terms to be hidden from view. It is one thing for biological life to be designated as the source of values in the sense that biological functioning appears to be a condition for the possibility of human agency that aims at values and makes validity claims, and quite another to argue that life itself *is* a valid value, and quite another yet to claim that it is the source of the normativity of values.[23]

The second stage of Rickert's case against the doctrine of *Sollen* in Life-philosophy began by recognizing that much of the power of Life-philosophy might remain even if its conceptual foundations were undermined. Life could still lay claim to value, even if this could not be deduced from science, biology, or through rational argument alone. It was on this point that Rickert—the theoretical philosopher of the "world-whole," the epistemologist and theorist of scientific concept-formation—started to show his true colors as a normative ethicist and transcendental philosopher of value (*Wertphilosoph*).

For Rickert, the problem with determining whether or not life itself is a value is that it leads to the question of what *kind* of life is good to lead, and this specification already leads beyond "mere" life into those values by virtue of which "mere life" first acquires ethical meaning. Rickert writes, "Certainly, life allows itself to be posited as a good, to which a value adheres [*anhaften*]. But often something else is hidden behind such valuations, which alone bears the value. Life itself then is not valued."[24] Rickert views the functioning of biological life as synonymous with vegetation; to value life as the *highest* end, then, would be to value bodily functioning or physical health above all else.[25] Yet the triviality of the truth that bodily functioning is the precondition and, in that sense, the source of human valuing, is reflected in the hollowness and even ethical banality of considering such functioning as the highest value and source of normativity. If life is to be redeemed as a value, it is only by redeeming other, autonomous, values that regulate it, orient it, and give it value.[26] Indeed, he argues that the list of values championed by the Life-philosophers themselves show this need to go beyond "mere" life: creativity, individuality, genius, honesty, nobility, aesthetic experience, youthfulness, joy, and freedom.

The third and related problem Rickert found with the practical implications of the concept of life is that it can lead to a quietist and superficial form of optimism.[27] Since everything happens through life, everything is the way it ought to be.[28] Rickert reads into Life-philosophy a disconcerting naturalistic form of theodicy that suggests that discontent with the status quo, or the state of nature, is simply a sign of decline among the biologically disgruntled or maladapted who are unable to affirm life. Just as the surreptitiously normative concept of Darwinian fitness could be used to champion the most diverse and conflicting political ideals from across the spectrum, so valuing life as the highest value led to a blank slate onto which all sorts of contradictory values could be projected. Indeed, Rickert goes so far as to claim that "[Life-philosophy] is then, so fiercely as its proponents fight against it, simply another of many forms of skepticism and nihilism."[29] Identifying Life-philosophy itself as nihilism—that is, as a philosophy for which any and all values were equally valid and therefore equally invalid—meant that it could not achieve the overcoming of nihilism, as Nietzsche originally had hoped. Because Nietzsche's goal was expressly to overcome nihilism by recovering nature's own valuing of itself, this charge brings the dramatic antagonism between these two philosophical perspectives into stark view. Of course, Nietzsche's affirmation was by no means superficial; indeed, it came from a genuine desire for life in the face of suffering that Nietzsche saw as the challenge of affirming a natural world shaped by the Darwinian "struggle for life." But Rickert's claim that this was a form of theodicy sought to draw attention to the fact that any reconciliation of the real and the ideal runs the risk of undermining efforts to intervene in the world so that the real more closely matches the ideal.

Knowledge for Its Own Sake

Rickert's criticism of the philosophical import of biology for questions of the world-whole, of knowledge, and of normative values constituted the bulk of his confrontation with Darwinism and the appropriation of evolutionary thought by the Life-philosophers. He saw his criticisms as establishing decisively the necessity of going beyond nature, life, and evolutionary dynamics when posing the central questions of philosophy, and therein lay the reason that philosophy had to depart from the sciences. It was the task of philosophy to aspire to knowledge of the world as a unified whole and to redeem the validity of those values that oriented human thought and agency, both within and outside of the scientific investigation. But, as we saw, the Life-philosophers not only presented competing doctrines in metaphysics, epistemology, and ethics, but also challenged this very notion of philosophy and its systematic task. Before showing how Rickert developed his own Idealist theory of value, we must address this last pillar of Life-philosophy. The rejection of systematic philosophy was an aspect of the antirationalist orientation of Life-philosophy, which saw reason to be a thin covering over more primordial natural drives, affects, and impulses. But Rickert set out to defend the Socratic and rationalist project against Nietzsche, whose original rejection of Socratic rationalism in *The Birth of Tragedy* set him on the philosophical path that led to Darwin.

Like Nietzsche, Overbeck, and Simmel, Rickert viewed science as a *task* that aimed at realizing a core value: namely, the intrinsic value of theoretical "truth for its own sake." He wrote that one engaged in scientific research "presupposes the intrinsic value of true thoughts and only through this becomes a theoretical researcher."[30]

Rickert addressed the implications of this position for understanding the natural-historical genealogies of both philosophy and science, with Nietzsche's counternarrative of the Socratic "theoretical man" clearly in mind. He makes, essentially, the same point that Simmel made in his late Life-philosophy about the autonomy of truth from vital-teleological aims:

> It took a long time until the value of truth in its purity came to awareness, and until people learned to "live" in it, in order to thus lend meaning to their merely vital life. In Greece, truth was first valued for the sake of itself, and the good, to which [truth] adhered, i.e.—science, was sought for the sake of truth. . . . One can seek and value [truth] in order to place it in the service of

> vital life. But then there is still no talk of "science." One studies only because one needs knowledge in order to live or for some other purpose. This is how it was everywhere at first, and it is still so for many. In Greece this relationship for the first time reversed itself. Human beings, also at first in only few exemplary figures, inquired [*forschen*] no longer in order to live, but lived in order to inquire. Through truth, life first became valuable. . . . Up to the present day, the theoretical person finds her unsurpassed model here.[31]

The various biologisms in their vitalist and mechanistic forms were unable to vindicate the validity of the guiding value of philosophy and science, the intrinsic value of truth for its own sake. Not only did this failure mean that this value could not be rationally redeemed, it also meant that suspicion was cast on the very possibility of such a value within the vital-teleological context. But Rickert thought that he had a philosophical ace up his sleeve in a style of transcendental argument inherited from Kant. These theories of value fell into a form of practical self-contradiction and Kantian heteronomy because they at once affirmed and denied the validity of a value at which their own philosophical efforts were aimed and upon which the validity of their own claims about life rested. The value of truth, of getting the world right, as a goal of intellectual inquiry, Rickert argued, could not be rationally rejected because such a rejection presupposed that very goal.

The consequence of Rickert's critique of Life-philosophy was a strict dualism between life, on the one hand, and valid values on the basis of which life was to be investigated, represented, and practically oriented, on the other. The final argument of Rickert's polemical work can help with the transition from Rickert's critique of naturalistic, Darwinian theories of values to his own alternative. He writes,

> There is no philosophy of life that does not presuppose truth about life and search for it, and this truth about life only lets itself be understood as something that is more than life. Thus, in the valid truths of life something other than life in its self-sufficiency is posited. The philosopher of life, of all people, cannot shake this other something, without thereby declaring his own philosophizing as meaningless. If Life-philosophy fashions itself as a doctrine of value, and so understands the essence of the timeless theoretical validity of value, then it must thereby break from the principle of the pure immanence of life. That is the philosophically significant point in connecting the problem of life to the problem of value.[32]

In order to comprehend the normative value that is implicit in every act of philosophical and scientific inquiry, a break with biology, evolution, and the monistic principle of life was necessary. Rickert sums up, "We can never stop asking about the 'meaning' [*Sinn*] of our lives, and only on the basis of values that are valid is [life] able to be interpreted."[33] The rift between *Sein* and *Sollen*, the *fact* of life and its *value*, is the basis of Rickert's attack on the messy blending of biological science and philosophy. Gaining clarity on this was the starting point and aim of his own approach to the problem of value, to which we now turn.

A Transcendental Theory of Value

Just as Hegel's *Phenomenology of Spirit* was conceived as an introductory ladder from the most basic empirical encounter with the world in sense perception to the standpoint of absolute idealism, so Rickert considered his skirmish with Life-philosophy to be a ground-clearing exercise for the rationalist and transcendental approach that he developed in his other writings. It was Kant, Rickert maintained throughout his career, who provided the original model for this idealist rationalism, because he first uncovered the *a priori*, but non-metaphysical, domain of normativity and strictly distinguished the causal order of nature from the normative domain of reason in both its practical and theoretical roles. Rickert first elaborated his Kantian, transcendental perspective on values in *Der Gegenstand der Erkenntnis* (The object of knowledge), which was published as a *Habilitationsschrift* in 1892 and then edited significantly through six editions as he further refined and defended his positions against critics.[34] *Der Gegenstand der Erkenntnis* was to be an introduction to transcendental philosophy, and in each of these editions he aimed to clarify this approach by defending it against objections.

The basic point of the book, elaborated in careful detail and through stepwise construction, is that knowledge claims necessarily presuppose normative criteria that distinguish valid from invalid thought.[35] The transcendental standpoint in epistemology could be understood by attending carefully to the normative element involved in every act of judgment, in life and in science. Every judgment, Rickert argued, involved bringing what he called the manifold "heterogeneous continuum" of sense experience under concepts.[36] He writes, "Even the most sober and cold understanding, with every step that it makes in knowledge, includes the recognition of values, and because of this all theory would lose its meaning (*Sinn*), if it did not presuppose a valid value (*Wertgeltung*) independent from the real act of valuing as its criterion or object."[37] Even basic empirical and descriptive claims, the determination of facts, instantiate criteria that distinguish *valid* acts of thinking from *invalid* ones. Recognition of this norm-laden character of perception, cognition, and scientific concept-formation was the first step in appreciating that knowledge itself was a normative concept and knowing was an act aimed at and regulated by a value.

Rickert argued that the act of judging *that* something is the case requires more than merely representing it, as if the mind were a passive mirror of the world. Representation too is an act of judgment that requires the formation of concepts, and in that act of judgment, a representation is always *affirmed* or *negated*. The value-laden character of cognition resides in acts of commitment that affirm or negate representations. These acts, undertaken by thinking agents, are at the basis of the representation of an independent, objective world. The key point is that what is affirmed or negated in such an act is not the representation as such, but the conformity of the representation to a norm or criterion. And, indeed, acts of affirmation and negation only make sense as instantiations of values that assume normative criteria. Rickert writes, "If knowledge is affirmation, its criterion is that which is affirmed, and what the act of judgment affirms or recognizes, lies always in the sphere of the ought [*das Sollen*], never in that of real being."[38] Rickert's transcendental epistemology aimed to show that the true "object of knowledge" (*Gegenstand der Erkenntis*)—what gives knowledge *objectivity*—is not the agreement between what is thought and a state of the mind-independent world. Rather, it is the agreement between a representation and epistemic norms that are taken to dictate what counts as *valid* cognition. The transcendental idealist "Copernican revolution" in epistemology originally ushered in by Kant, Rickert held, is the recognition that knowledge of an objective world consists in the agreement between representations and *a priori* norms, not between representations and the mind-independent world.

Rickert criticized the positivist and scientistic conception of knowledge as value-free because it assumed a faulty conception of representation as a "copy" (*Abbild*) of the external world in thought, and so a faulty conception of mind as a passive mirror.[39] Getting rid of this correspondence and pictorial theory opens up the space for seeing knowledge as a norm-guided act of thinking in which the manifold content of sense is taken up into conformity with norms. Science was a *task* with an *aim* in a few senses. First, it was not just a free-floating body of information about causal relations between physical, mind-independent entities. Second, science was to be placed within the sphere of human action and the *teleological* structure of human rational agency, as it guided itself toward the realization of values recognized to be valid. Of course, Rickert's argument debunked the notion that purely theoretical inquiry is or can be genuinely value-free, and on this point he and the Nietzscheans agreed. But, the crucial point of disagreement was that the norms and values involved in cognition were not merely biologically necessary in that they were caused by basic teleological drives common to all living things. Instead, they were rationally, and transcendentally necessary in that they were presupposed in and through reason's self-legislated aims, which included the aim of knowing the objective, mind-independent world. As Rickert wrote in his early epistemological work, "A transcendental ought as the object of knowledge is, regardless of what epistemological standpoint one assumes, indubitable, because it is the presupposition of every true judgment, indeed even every theoretical doubt and thereby the presupposition of every standpoint, with the inclusion of skepticism."[40] Philosophy as the theory of value thus had an indispensable role to play next to the sciences in that it provided a theory of validity presupposed by all thinking and acting, including the constitutive norms and values of scientific investigation.[41]

In 1921, directly after his critique of Life-philosophy, Rickert published the first volume of his comprehensive *System of Philosophy* (*System der Philosophie*). This work began by offering illuminating reflection on the concept of value, which, as we have seen, is basic to his characterization of human agency. Rickert argues here that *value* is a theoretically primitive term that covers a collection of phenomena—norms, laws, criteria, goals, rules and ideals—all of which shared the unique feature that they regulated and oriented human thought and action, and that they explicitly contained validity claims. This class of phenomena did not refer to properties of the world as it is, but to how human thought and action "ought" to be oriented and guided in order to realize its aims. Like Simmel, whose vocabulary was influenced by these late nineteenth-century, neo-Kantian debates, Rickert circumscribed the realm of values with the broad term *das Sollen* ("the ought"). Just as, in the realm of being, all entities must be categorized either as "real" or "not real," in the realm of the "ought," all values must be categorized as either "valid" or "invalid." Rickert wrote, "It is above all necessary that we keep our eyes open for the 'other' world of meaning [*Sinn*] and value, that we do not place it in a mysterious beyond but recognize that it is very near to us. We 'experience' non-real values just as immediately as real existing objects; indeed, the non-real plays a decisive role in our waking life. We do not 'live'—that is, we do not think, will, and act—without values, which give direction to our thought, will, and agency."[42]

The basic error in the theory of value made by Rickert's contemporaries, he claimed, was to confuse the realms of being and value by consistently regarding values as identical with the beings to which they adhered (*anhaften*). Life-philosophy too committed this error. But values were not to be placed in things or identified even with the *acts* of valuing that go on in any individual agent's psyche. Instead, they were presuppositions of acts of valuing, goals at which acts of will and thought aimed. To say that values are "non-real" (*Irreal*) did not mean to suggest that they are subordinate to or less important than what is; rather, it was to argue that applying the categories of "existence" and "nonexistence" to values was misguided. Indeed, raising the category of reality above that of the non-real was already to invoke a value.

For Rickert, value thus came to define a sphere occupied by specifically human thought and action that was utterly basic and distinct from the mind-independent world that the sciences aimed to grasp in concepts: "What, finally, a value is in and of itself of course does not let itself be 'defined' in a strict sense. But this is precisely because it indeed deals with one of the final and non-derivable concepts with which we think the world, and the concept of value shares this deficiency, if one will call it that, with the concepts of being, existence, and reality [*Realität*] or actuality [*Wirklichkeit*]."[43] Rickert goes on to define value in relation to being: "The essence of what exists, which we do not mean with the term *value*, lies in this, that what exists does not 'concern' us, if it only exists. It leaves us 'indifferent' and does not 'move' us. We represent it. It is just there. . . . In contrast, a value that comes to our attention as a value never leaves us indifferent in this way. . . . We take a stance toward it, are 'involved' or 'interested' in it, feel ourselves moved and *called in our spontaneity*, are bound by it, we do not simply represent it."[44]

The term *value* invoked the claim that the world makes on desire, interest, thought, and action. Value was a primitive and irreducible concept because humans did not merely mirror the world in a disinterested manner. All judgment involved the recognition of "oughts," and these "oughts" were, finally, to be interpreted in terms of which values made an irrefutable, binding claim on self-reflective human agents. The application of the category of validity to causal events that simply occur in nature was nonsense. The only sphere in which this term could be meaningfully applied was that of rational, understanding agency. Again, Rickert's charge was not simply that naturalistic and positivist thinkers, like the Nietzscheans, committed a logical, naturalistic fallacy by deriving an "ought" from an "is"; it was that they had mischaracterized the form of agency and the capacity for representation that humans possess. Moreover, this form of agency was presupposed by the very claims to validity that humans made for their own actions and in critical debate with one another. Indeed, this form of agency, which made the acts of cognition and will that were involved in science possible, thereby generated the possibility of a naturalistic fallacy in the first place.

Science and the Meaning of Life

Even in his early writings, Rickert was aware that his notion of philosophy as the theory of valid values had wider implications outside of epistemology. Rickert shared with the Life-philosophers the idea that science was a task that belongs within a wider theory of human life. Because philosophy could now rightly be understood to be explicating the unique *rational-teleological*, as opposed to vital-teleological, structure of human agency, it now seemed to be in a position to recover another of its most ancient tasks. This was not the epistemological goal of clarifying the normative character of knowledge, nor the scientific goal of causally explaining natural events. Instead, it was the existential task of providing orientation with respect to the "meaning of life" (*Sinn des Lebens*) as a whole.[45] In an early essay from 1910, "Vom Begriff der Philosophie" (On the concept of philosophy), Rickert articulates the problem of the meaning of life in existential terms as the problem of acquiring a worldview that yields a practical orientation: "What do we actually mean, when we talk about a 'worldview'? Through this [idea], we do not in fact just want to come to know the causes that produce us and all other things, and so to explain everything in its causal necessity, but instead we also want to win an understanding of the world that, as one tends to say, teaches us the 'meaning' of our lives, the significance of the self in the world. . . . What are we actually striving towards? What is the goal of existence? What should we do?"[46]

The problem associated with acquiring a worldview was the problem of the meaning and value of life that the Life-philosophers regarded as central to the world's religions.[47] However, for Rickert, the question of the meaning of life could only arise for thinking agents and so was radically independent of reflection on nonhuman life, biological processes, or evolution.

Rickert writes in this essay that "when we yearn for a worldview that tells us what the world means, there we are asking—when we understand ourselves correctly—whether our life has *value*, and what we ought to do in order for it to become valuable."[48] Or, further, "The validity of values remains in all situations primary for the question of the meaning of life."[49] We have seen that Rickert's epistemology was aimed at showing the rational necessity of valuing truth *for its own sake*, and this was the constitutive value of scientific inquiry. Thus Rickert could show that philosophy and science were relevant to the ancient concerns over an orientation to the meaning (*Sinn*) of life by demonstrating the validity of this intrinsic value.[50] However, an obvious problem arises here for Rickert's project. He might have defeated biolog*ism* (and psycholog*ism*) in the theory of knowledge by offering a transcendental argument for truth as a valid value. This was good enough to show that at least one value is rationally and transcendentally valid; that is, the value of truth did not depend upon whether any humans in fact recognized it or were moved by it at all. But how could he offer this kind of a transcendental argument for the values in other areas of cultural life? Would all goals outside of truth, and all activities outside of science, have to succumb to the irrationalism and biologism that he so struggled to oppose?

Rickert called the values outside of science "nontheoretical" values and listed among them the constitutive aims of ethics, aesthetics, and religion. This list obviously carries Platonic resonance in relation to the transcendental values of the Good, the Beautiful, and the True. Each of these spheres, Rickert claimed, posited ideals that were taken to be intrinsically, non-instrumentally, valid. Moreover, each of these arenas posited values that showed discontinuity among human values and the merely biological needs that humans shared with other living things. Always careful to note the values presupposed by intellectual work, Rickert reminds his readers that providing a theory of such nontheoretical values is itself not a value-free task. As a theoretical project, it presupposed and aimed at the same theoretical value as science—truth and theoretical clarity *for their own sakes*. Crucially, Rickert argues that theoretical starting and ending points limit what philosophy and science can say about nontheoretical values. Rickert writes, "Whoever demands indubitable objective values, in the sense that their validity, independent of every subject, allows itself to be proven or rationally justified, must not only begin with the domain of theoretical values, but also cannot in a certain sense wish to go beyond it. The objective validity of ethical, aesthetic, religious, or other nontheoretical cultural values eludes every scientific proof. This can be easily understood from the essence and peculiarity of nontheoretical values. If their validity were theoretically justified, then they would cease to be nontheoretically valid."[51]

Rickert's remark here was directed primarily against the error of what he called intellectualism, which was also a target of Simmel's critique. The error of intellectualism was to take the transcendental argument for the validity of the theoretical value of truth as a demonstration that truth is the *only* valid value or the *highest* value, to which the other nontheoretical spheres were subordinate. Rickert intended his project of a system of values to respond precisely to this problem of how to think about the validity of nontheoretical values without subordinating them to science. Like Simmel, Rickert thought that the aim of a system of values was simply to clarify the guiding values that constituted each value sphere, supplying its internal criteria of justification and critique. The aim was not, and could not be, to rank these diverse values.

Rickert's placement of philosophy and science within the sphere of human agency was the first step toward a general Kantian theory of culture, which he developed in late writings that followed his critique of Life-philosophy.[52] This theory of culture was to provide a bridge between the philosophy of science and epistemology and broader cultural concerns, including the relation between science and the existential interest in a worldview that could orient us with regard to questions of meaning and value. Rickert tellingly used the term "care" (*Pflege*) as a synonym for "culture" because it pointed to the fact that culture is simply the product of the goal-oriented, *teleological* nature of human agency, which is invested in realizing and sustaining values.[53] Thus, science, ethics, art, politics, and religion all belonged to a general theory of culture as interested and active care for the maintenance and realization of values that pertained to all arenas of individual, social, and political life.

Instead of transcendentally grounding nontheoretical values, Rickert's idea of a system of values set itself the more modest task of showing simply that the values aimed at in various spheres of culture were *rationally necessary*, and not "merely" biologically or historically given. The way to defeat naturalistic and historicist worries that, in his view, undermined the validity of these values was to show that they could be derived from the structure of human agency. Only on the basis of such formally derived principles could one look to the material of cultural history or nature to judge which values found there could make a defensible claim to being genuinely and universally valid, independently of the given fact of their being valued, in the way that the value of truth was shown to be through transcendental argument. The theoretical challenge of a system of values was thus to marry a normative theory of practical rationality with the messiness and particularity of both cultural *and* natural history.

The key principle of Rickert's theory is the principle of "completion" (*Vollendung*). Rationally valid values, he argues, offer *completion* and *fulfillment* of some basic human capacity. The postulate of forms of completion and fulfillment was itself evidence of the emergence of *rational* teleology into the sphere of merely biological or vital teleology. Rickert hoped to show that the content of nontheoretical and theoretical cultural values in the spheres of science, politics, ethics, aesthetics, and religion could be shown to be conceptions of what it takes to complete or fulfill some universal capacity of human life. Capacities of intuition, of feeling, of social interaction and bonding, of practical reason, of theoretical reason, and even of personal bonding in friendship could all seek a form of completion. "Intellectualism," "aestheticism," "moralism," "eudaimonism," and "religion" were all worldviews that resulted when the respective completion of any arena was taken to be the highest aim, to which all others were subordinated.

All actions, Rickert claimed, aim either at a future, never fully realizable goal called an "infinite totality" (*un-endlichen Totalität*); a satisfaction in a distinct and fleeting moment called a "realized particular" (*vollendlichen Partikularität*); or, finally, a fully "realized totality" (*voll-endlichen Totalität*) in which a goal has been achieved once and for all.[54] Moreover, Rickert claimed that these possibilities for the realization of values were exhaustive and exclusive, and they allowed his scheme of values to yield a preliminary but not conclusive *ranking* of values. Even if a scientific philosophy could not determine exactly which values are indeed realizable in nature at all, the principle of realization allows a rank ordering according to the magnitude of the value that is aimed at in each sphere. The "infinite totality" that seeks a future value that can never be fully realized stands subordinate to the "realized particular," and both are subordinate in value to the fully "realized totality," which is the highest form of realization that an ideal can achieve. One can see in the end that Rickert's philosophical answer to the ancient question of the meaning of life was not to show which one of these values is in fact the highest, but to show that the validity claims made by each domain could be redeemed because they offered the completion of some basic human capacity and sphere of activity. The ideal of completion was the aim of reason, not of mere life.

Religion in the System of Values

Rickert's philosophy of religion was an attempt to relate the goals posited and striven after in religions to the formal, rational principle of completion upon which the validity of nontheoretical values was based. Rickert called his approach to religion an "interpretation of the meaning of religious life" (*Sinndeutung des Religiösen Lebens*).[55] Unlike the Life-philosophers, who sought to show that religious values like the ascetic ideal arose out of basic natural drives and affects that were continuous with biological life, Rickert wanted to show that religious values too could be derived from the principle of completion that reason introduced into the world of biological agency. Rickert identifies the constitutive value of religion at various points as "holiness," "perfection," "divinity," and, deliberately referring to Nietzsche, as a "superhuman" (*übermenschlich*) state of fulfillment.[56] While science aims at truth, ethical life at moral virtues, political life at social harmony, and aesthetic life at an intuition and enjoyment of the senses, religions posit and affirm holiness and human perfection as values to be realized in the world and against which both humanity and the world are measured.

Rickert's understanding of perfection shows important similarities with Nietzsche's and Overbeck's focus on otherworldly asceticism and Simmel's conception of transcendence as the characteristic aims of religious life, the aims that gave religions an intrinsic criterion of justification, critique, and estimation of success and failure. For Rickert, the religious ideal is *übermenschlich* in that it is precisely targeted at transcending the limitations of the human perspective. In his scheme of completion, religions posit and strive for "fully realized totalities," final consummations and realizations of all human striving. However, despite the rational necessity of positing such an ideal, Rickert writes, "Once we understand ourselves, we can never wish to realize the religious ideal. This intention includes the will to become a god. It lies in the concept of the holy as something super-human [*übermenschliche*], an absolute perfection, whose realization looms over us as unachievable, on the one hand, but at the same time is valid as an avoidable ideal against which we measure all human work [*Menschenwerk*], on the other."[57]

The ascetic nature of religious values, in Rickert's estimation, stemmed from their offer of a form of completion whose potential realization in the mundane, earthly, and natural world appeared impossible. It is for this reason that religions posited different orders of reality beyond the mundane. Rickert even sees Nietzsche's ideal of unconditional life-affirmation as just such a "perfectionist" impossibility.[58] The fact that religious life posits the transcendence of the everyday and "human, all-too-human" as its ideal leads to inner division through consciousness of impossibility, powerlessness (*Ohnmacht*), and distance from the "holy" and "divine" that it takes as its aim.[59] Yet, because it holds onto and refers all values to a final completion and fulfillment of all human action and thought—that is, a full and final completion of life once and for all—its normative claim to rational validity could not be dismissed. Indeed, Rickert suggests in wild contrast to conventional wisdom regarding the rationality of religion today that this principle of final completion is the *most* rational principle because it offers total finality, an ultimate fulfillment of nature and reason alike.

There is another aspect of the religious ideal that gives it a unique philosophical significance. Affirming the normative validity of the religious ideal, a rational postulate of full and final fulfillment, presupposes the belief in its possibility and thus of the power of this ideal over the real world. Indeed, Rickert defines belief in God precisely as belief in the "power of the ideal over the real."[60] Religions are special in the chorus of cultural values because they posit "value-realities" (*Wertrealität*) that affirm the ultimate unity of the ideal and the real, valid values and the existing world. To believe in this ultimate unity, and in the power of the ideal over the real, was to believe that nature was such that the "ought" of full completion implied "can." Because of this, the core claims of religious worldviews ventured into the domain of metaphysics closed off to a critical philosophy limited by reason, for which a unification of the rift between *Sein* and *Sollen*, value and existence, remained inconceivable.[61] Rickert's philosophy of religion, which again is merely an *interpretation* of the meaning of religious life (*Sinndeutung*), need not commit itself to pursuing the religious ideal, nor to the affirmation of the power of the ideal over the real, nor to the actual existence of this ultimate unity of value and reality.[62] However, the underlying logic of religious life nevertheless involved the idea of a value-reality that guaranteed the possibility of a final completion of life, and this idea was in fundamental conflict with a conception of the order of nature as value-neutral.

Conclusion

At its heart, Rickert's epistemology and theory of values carried out a revival of Kantian Idealist philosophy amid the competing schools of late nineteenth-century German thought by insisting on the priority of *Sollen* to *Sein*, value to fact, the *ideal* to the *real*. His transcendental turn restricted the domain of the positive sciences by opening up investigation into an area that—he argued—was logically prior to and presupposed by them and was indeed radically autonomous from the existing world of nature that the sciences were to describe and explain. He saw Kant's most important discovery to be the discovery of the realm of non-real, but rationally binding, normative values. Recognizing this realm was crucial for understanding what was at stake philosophically—and even existentially—in the encroachment of the ever-growing natural and human sciences and the rise of irrationalist and purely instrumental notions of value in Darwinian and naturalistic philosophy, including American Pragmatism. However, this project created a fundamental and insurmountable dualism. Instead of establishing a dualism of mind and matter, thought and extension, the in-itself and experience, language and reality, Rickert argued that the fundamental dualism that philosophy needed to confront was that between nature and the *non-real* realm of valid values that any objective *knowledge of* nature implicitly affirmed and presupposed.

Throughout his career, critics attacked Rickert's strict dualism between real *Sein* and non-real *Sollen*, arguing that a more fundamental ontological unity had to be sought.[63] As the generation of post-Kantian German Idealists criticized Kant's dualism, so Rickert's dualism inspired severe criticisms from his contemporaries that are echoed by philosophers today.[64] Values could not be left untethered to nature or any other domain of reality. Just as troubling, though, was that Rickert's strict distinction between biological and rational teleology left it highly questionable whether the values aimed at in science and cultural life were realizable in the natural world at all. If normative validity was wholly autonomous from life, then how could the evolving Darwinian world be a world within which these rationally necessary values could be realized? As Nietzsche, Overbeck, and Simmel also worried, nature seemed not just indifferent, but even resistant toward cultural values that aimed beyond life's own ends. Rickert's defense of the autonomy of reason from nature seemed to leave all cultural ideals in the same position as the religious ideal: Humans were rationally driven to work toward goals whose very possibility could not be rationally established. The challenge of the Darwinian view of nature was whether or not "ought" indeed implied "can," whether it was right to think that this evolving world of nature was one in which the aims of reason (and religion) could be realized.

In his late essays, Rickert repeatedly returned to this paradox of human, rational agency.[65] If philosophy as a theory of valid values could not help to justify the sense that nature as a whole is a world wherein the necessary aims of thought and culture can be realized, he worried that the "irrationalism" that he all along sought to avoid might return. In addressing the problem of the final unity of culture and nature, of *Sollen* and *Sein*, Rickert turned to the realm of *Glauben* (faith) and its relation to scientific and philosophical *Wissen* (knowledge). The only way that *Sein* and *Sollen* could be reconciled, he argued, was if there were indeed a final "value-reality" (*Wertwirklichkeit*), an underlying metaphysical unity of value and reality. But to postulate such a reality transgressed the dualism central to Rickert's theory of value, and it transgressed the strictures of "self-reliant" philosophy that could rationally redeem its basic assumptions.[66] Rickert writes, "In the concept [of a "value-reality"], the separation of value and reality, which critical thought demands in connection with the critical separation of theoretical and practical reason, is no longer to be maintained. The valuable itself must be effective and in this respect must be 'actual.'"[67] Since such a unity could not be conceived theoretically, literally it could not be thought; only a foundational, basic faith in the possibility of realizing valid values in the world could make sense of the undeniable normative demand of such values on agency. While a metaphysical value-reality could not be conceptually grasped, it was existentially necessary to postulate that the aims of reason were indeed "in harmony" with nature and, therefore, at home in the Darwinian world.

Rickert thus finally came to affirm that there is a basic, extra-rational faith that is presupposed by the exercise of human agency. Rickert writes, "Thus nothing remains, if we wish to orient ourselves positively to the problem of a value-reality, than to *transgress* our knowledge. We have to *believe* in a realization of values in the world. . . . Without such a faith, our orientation to values would also lose its meaning, for nothing would validate that it actually has an effect in the world."[68] This faith affirmed the metaphysical unity of nature and value, the ideal and the real, whose separation, if we recall from the introduction, marked the way in which thinkers in the post-Kantian, post-Hegelian generation outlined in these chapters approached foundational questions concerning the relationship between ethics, metaphysics, and science. In Rickert's final assessment, neither philosophy nor science could make sense of how cultural values that were rationally valid could indeed be realized in the world of Darwinian evolution.[69]

Rickert's philosophy as a whole shows that his confrontation with Darwinism and Life-philosophy penetrated into the fundamental problems of agency, value, mind, metaphysics, and knowledge that have shaped Western philosophy. In the end, Rickert's philosophy of religion turned to the post-Kantian German Idealist J. G. Fichte (1762–1814), on whom Rickert wrote an early essay dealing with the heated controversy over Fichte's alleged atheism that led to his dismissal from the University of Jena in 1799.[70] Like Fichte, Rickert finally argued that only a scientifically and rationally ungrounded, basic faith in the power of the ideal over the real, a philosophically refined and rationally purified form of religiosity, could overcome the skeptical and nihilistic conclusions that faced so many philosophers in Germany who were confronted with the philosophical consequences of Darwin and the picture of the world being generated by the natural sciences. Whether this faith reintroduced the possibility that Nietzsche's life, hungering for itself, was at the basis of the ideal is not a possibility that Rickert entertained. But, like the Life-philosophers, Rickert too shows that the problems that Darwin and the late nineteenth-century biologization of philosophy posed went well beyond epistemology and the philosophy of science. Darwin's impact bled into the extra-scientific, existential concerns that, at least for Rickert, only a return to metaphysics appeared able to answer affirmatively.

Conclusion

The previous chapters presented varied philosophical responses to evolutionary biology in the wake of Darwin that began with Nietzsche's use of a conception of life to confront the problem of nihilism. Each chapter showed how understandings of life that appeared to be validated by the creativity demonstrated in the process of evolution were used within a systematic approach to questions about the sources of value, the origins and nature of religion, and of what science as a human practice is. Reflection on the relationship between life and value became the link that coupled the natural sciences to philosophical questions in meta-ethics and metaphysics that were also core elements of religious systems of belief and practice. But underneath these more abstract and theoretical questions, there was another philosophical urgency that these thinkers brought to their reflection on evolution. This was the question of what human ideals were capable of being realized and what forms of human fulfillment were possible in a Darwinian natural world. They brought to the life sciences an ethical and existential seriousness about the human quest to realize ideals in the natural world and about understanding what forms of spiritual satisfaction and achievement were possible. In effect, they looked to biology to understand the relationship between the real and the ideal, and biology in particular was important for this question because processes of life were fundamentally appetitive—they were *after* something.

For Nietzsche, all forms of life shared the fundamental appetite whose *telos* was life itself. This teleological character gave Nietzsche resources to reconceive life as the creative source of all of the value-laden ideals embedded in human culture. On the one hand, Nietzsche's concept of life was a descriptive concept that was meant to capture what living things are after, what they want, need, and seek. This, of course, was to help him understand what humans most fundamentally seek and need as well, which found expression in the ultimate aims of religious and ethical systems. On the other hand, Nietzsche used life's nature as an evaluative criterion, one that could diagnose and revalue the constitutive values of both religious life and scientific investigation. Life itself was a natural value and a natural phenomenon to be valued. Thus, valuing life became a criterion of health and flourishing. The fundamental tension that remained in Nietzsche's project was the extent to which his conception of life could justify the normative claim that valuing something *other* or *beyond* life, against which life itself could be seen to be flawed, was a form of defect and, in his words, "sickness." For Nietzsche, the notion of validity became synonymous with vitality, and recovering nature's own self-affirmation became the ultimate biological and ethical achievement.[1]

For Overbeck, the lesson of biological life and evolution was different. Nature's creative process was not the sign of a fundamental, ultimate aim shared by all forms of life, but of finitude and the inevitable cycles of youth, maturity, and old age that even cultural forms like science and religion were subject to. Vital affirmation was a symptom of the naive youthfulness and unreflective self-confidence that appeared during the early stages of new cultural formations but inevitably also faded and degenerated through time as new forms of life in their youth emerged. The vital drives that produced myths and erotically sustained religious stances toward the value of life too went through life-cycle stages in their historical development. Moreover, the aim of this youthful vital drive, oriented around the question of life's own value, stood in stark conflict with the aim of science to arrive at purely theoretical knowledge *for its own sake*, disconnected from the question of the value and meaningfulness of life. For Overbeck, a proper understanding of the dynamics of the temporal pattern of life-in-history undermined the possibility of deriving general criteria of value, even of the value of life itself. There was no constant *telos* in nature's vital cycles of birth, development, and decline, and so the values embedded in both science and religion could claim only a deeply historically and vitally conditioned form of validity.

Like Overbeck, Simmel rejected Nietzsche's notion of life as the abiding and ultimate *telos* within the creative flux of evolution and its generation of organic and cultural forms. Life was the source of values and the affective basis of their normative grip on human thought and action, but in and through its own creativity, it became more than life. In self-conscious human agency, drawn toward ideals, the creative flux of life manifested itself in a novel form, that of objectively valid values confronting rational agents who were to choose among them. These values were encountered in experience, not as products of subjective preference or appetite, but as given facts belonging to a genuinely independent and authoritative order that pulled individuals toward future realizations of themselves and of the world. This was the undeniable phenomenology of value and human agency, which these intrinsic values made claims on. But this phenomenology too had its own evolutionary history for Simmel in social and economic interaction and even, finally, in the vitality of life's creative drive; these intrinsically valid values were only momentary eddies within an underlying dynamic, Heraclitean, flux. As in Overbeck, the validity of values still depended on the active pulse of the creative drive that produced them in the first place, and it could never achieve full autonomy from this. But unlike Nietzsche, Simmel did not see life as ceaselessly thirsting after *itself*. It generated and strove for more than life, for fixity and relative permanence in forms that transcended its underlying, dynamic character. Simmel went beyond Nietzsche to account for the emergence of cultural values—such as those found in religions and in science—that were autonomous from merely vital-teleological purposes. In the course of social evolution, these established their own unique goals and created new final aims, for the achievement of which individuals, and life itself, were to become the means.

Rickert was right that each of these thinkers rejected materialist understandings of life and Darwinian evolution as effects of mechanical interactions between aimless physical substrates. Nature was a creative drive and power, and evolution became a clear example of the creative pulse of nature toward new and diverse forms. The need for this antimaterialist notion of nature's creativity arose precisely because cultural aspirations embedded in science and religion could not be reduced to material desires and interests, and the only way to understand how these spiritualized needs and desires could emerge from nature was to see life itself generally as an appetitive drive and not merely a blind mechanism. Like the organic forms that belonged to biology proper, the values that humans came to seek and desire were products of the goal-directed, teleological character of biological life. The Nietzschean appropriation of Darwin presented here revolved around the question of what kind of teleology there is in the biological world, and so upon what natural basis the validity of human values rested. This was the background necessary to show how both religious interests in a way of life in harmony with reality and the scientific interest in truth *for its own sake* were intelligible products of biological life.

Against this vitalist philosophy, and against the encroachment of Darwinian thinking on philosophy in general, Rickert argued that if any teleology and value could be drawn from biological life alone, it could only be the value of mere bodily functioning. Moreover, the Nietzscheans had crucially confused investigation of the physical and biological conditions that made values possible with investigation of their validity. The question of normatively valid values could only arise and indeed make sense in relation to regulating the conduct of rational, self-reflective, and understanding agents. In other words, only reason, with its capacity to set its own ends that seek a completion of the various natural capacities that humans are endowed with, could be a source of normativity. This was typified in Rickert's treatment of the concept of "health." It was one thing to use this term to describe different states of the functioning of a biological system. But the term *health* indicated that certain states were *to be cared for* and *to be cultivated* over others. This term could only have meaning in relation to the conduct of a rational, indeed also moral, agent who was capable of favoring some states of affairs over others and of intervening in the world to produce them.

If there were *valid* values, then, philosophy had to go completely beyond biological life in order to account for and vindicate them. This is why Rickert was so worried that the turn to biology could only result in a form of nihilism—the sense that life was not ultimately valuable—that Nietzsche too wanted to challenge and overcome. Nietzsche pleaded for a stance that was "sure of life and its ideals," but this, Rickert argued, could not come from reflection on biological drives, desires, or functions. The value of these drives and desires could only be vindicated by being taken up in the form of rational agency, as means toward its seeking of completion. The ultimate and intrinsic values were not "vital values" (*Lebenswerte*) like bodily health; they were what Rickert called "cultural values" (*Kulturwerte*), and these were values that transcended mere bodily functioning and that rationality allows us to recognize as centers of their own intrinsic worth. These were the values enshrined in the spheres of art, science, ethics, law, and religious practice. The price of Rickert's critique was a strict dualism between nature and value, and this is something that many philosophers both in his time and today have resisted.

It is tempting to want a sort of Hegelian *aufhebung* that would show the partial truth of each of these positions and reveal a middle way that preserves them all. Perhaps the Life-philosophers failed to appreciate the unique character of rational, norm-guided agency within the sphere of biological life. And Rickert perhaps failed to appreciate that norm-guided agency too was a form of life, continuous with and perhaps even unintelligible outside of the sphere of biological activity from which it emerged. Valuing life could not be dismissed as valuing *merely* vegetative bodily functioning, because this functioning was already incipiently aimed at maintaining the organism and ensuring both the exercise of its capacities and the achievement of its minimally biological aims. Moreover, human capacities of rationality, mind, and desire were in and of life's evolution. In viewing life as mere vegetation and bodily functioning, Rickert adopted the mechanical view of biological life that perhaps did not do justice to Nietzsche's claim that living activity is fundamentally after something or to the creativity and novelty in the evolutionary process that so impressed Overbeck and Simmel. For the Life-philosophers, human agency and valuing was thus continuous with nature, whereas for Rickert, it could not be so, precisely because it had discovered *valid* values that went beyond merely biological needs. For this reason, understanding the aims of science, of philosophy, and of religions meant leaving the rest of nature behind. While an *aufhebung* is desirable, it is still a task that is left for contemporary philosophers to achieve.

I have also suggested, more controversially, that Nietzsche's turn to the life sciences and to life itself was driven by an interest that is best regarded as religious. Evolution and biology were probed for the sake of questions about what to seek in life, what values to strive to realize, or—indeed, even more religiously—what sort of human life was most in harmony with the most fundamental dynamics of evolution that gave rise to us. Life became the subject of an existential hermeneutic—that is, an inquiry into the *meaning* of nature, its significance for how one lives, and whether or not human life, oriented by the recognition of ultimate values and by the search for completion, can find satisfaction in the world of nature. These were questions of what the Life-philosophers themselves called *vital* concern, but they raised a host of philosophical puzzles of how this concern itself emerged in nature and whether or not it too made sense in relation to biological drives of survival and reproduction. What point was there for a product of evolution to worry about the meaning of its existence, or whether or not it was living in harmony with the universe? While other living things lived instinctively, humans were in the position of having to live with a reflective orientation toward values and with the burden of making choices about which life is best. The intellectual efforts of the Life-philosophers were a manifestation of this task and the burden it placed on us.

One of these burdens was precisely the human capacity to *justify* values rationally and the burden of feeling the need to do so. For Nietzsche and Simmel in particular, this capacity and the burdens that came with it arose through social struggle and the dynamics of power that operated between individuals and groups. Further, the need to rationally justify values was itself the origin of conceptions of transcendent realities that redeemed values with reference to a reality beyond nature. But Nietzsche aimed to satisfy this need and to supply this justification through life alone. For Nietzsche, valuing life was at bottom a biological necessity; for Simmel and Overbeck, it was not even that; and for Rickert, it was a rational one derived from the validity of other values. In rejecting life as a naturally valid value and criterion of value, Nietzsche's critics also rejected his religion of life-affirmation.

These thinkers and their disagreements show that Darwinism in Germany was plunged into a dense and difficult set of problems surrounding normative validity and value, idealism and materialism, mechanism and teleology, religion and science that defined the post-Kantian and post-Hegelian traditions in German thought. Of course, these debates also revisited foundational issues that date back to the sources of Western philosophy and religion. The focus of these debates was on meta-ethical and metaphysical questions about the sources of values, their normative status, and the possibility of their realization in nature, but these questions now faced the new and unavoidable reality of evolutionary drives, the deep history of life, and the creative power of the living world. These thinkers are examples of a range of philosophical interests in Darwin and evolution in this period that went beyond what these thinkers, and what we today, might regard as purely scientific—if this is defined in their terms as interest in acquiring knowledge *for its own sake*. These thinkers were devoted to discovering the metaphysical and moral meaning of the evolving world of life.

Value, Teleology, and the Representation of Life

There are many points to draw from these debates that might be of lasting interest to contemporary philosophers working on these issues. One lies in a middle ground shared by the Life-philosophers and Rickert. This was the point that representing life and inquiring into its teleological nature is not a strictly scientific activity, answerable to purely theoretical values and aims, but also a moral, aesthetic, and even religious activity. As Rickert argued that an "ought" and an affirmation of value lie at the basis of every act of judgment, so too did the Life-philosophers understand representing the world to be a value-laden act. In so doing, these thinkers recast the relationship between science and other spheres in terms of their ultimate values and aims. So, while the scientific aim was to represent the organization of life independent of any ethical, aesthetic, or religious goals, the question of what life is, and its relation to how things are metaphysically with the universe as a whole, was a question of which values were valid and which were, indeed, the highest or ultimate values.

Let me try to put this point another way. By viewing the representation of nature as an activity normatively guided by commitments to values, these thinkers recast how we might understand conflicts between different conceptions of life. A picture of the world might be generated in accordance with the guiding aims and internal criteria of any of the distinct spheres of value, each with its guiding ultimate value. However, the claim that the picture of the world that held final authority was the one generated in accordance with the scientific value of purely theoretical truth *for its own sake* was not a scientific claim about the world-in-itself. Instead, these thinkers insisted that this was a claim about the relative validity of purely theoretical values in contrast to other values and about the *ranking* of the value of purely theoretical truth next to the core practical, aesthetic, or existential values that shaped the internal norms of other domains. But the procedure of ranking values could only be done on the basis of the internal norms proper to one of these spheres; there was no master norm or value to which one could appeal. The representation of life, in particular the teleological character of organisms or of evolution as a whole, as *after* something, then, was an act that was fundamentally tied to the adoption of a worldview, which meant a commitment to a stance regarding which sphere was indeed ultimate. This decision about which sphere was highest was a decision about which sphere of the ideal, and which human capacities that gave rise to these spheres (again, understood here as broadly ethical, aesthetic, religious, or theoretical), were the most trustworthy windows onto the real. Moreover, it was also a choice about which human capacities and ideals led to genuine flourishing.

This approach to the representation of life, and to the representation of nature as a whole, has intriguing results when applied to the question of teleology. It challenges us with the claim that representing life is an act loaded with our own commitments to ideals and values—one that purely theoretical inquiry *for its own sake* may not have sole or ultimate jurisdiction over. It is an act that we may not be able to quarantine from nonscientific values. Representing life and determining its teleological character hang in between the spheres of science, aesthetics, ethics, and religion. Considered as a product of cause-and-effect relations, embedded within a wider context of nature considered as a sum total of cause-and-effect relations, life appears as a mechanical effect and mechanical cause that is indifferent to values. But regarded as an object of aesthetic pleasure and sublimity; or as an object of moral concern, integrity and respect; or as a sign of transcendent value; or even as a phenomenon *to be affirmed* and consented to, these cause-and-effect relations can take on the character of ends, both in themselves and for humans to incorporate into practical life. Each of these spheres answers to its own intrinsic norms and criteria of justification derived from a guiding and final value that individual agents are committed to realizing.

This picture of representation opens up the possibility of representing whether there are ends in life, and what they are, in different ways, each valid within its own sphere. The Life-philosophers and Rickert challenge us with the perhaps unsatisfying conclusion that these spheres are not, finally, rationally adjudicable according to criteria outside of those embedded within each sphere. They also challenge us with the assertion that the norms of rationality too are internal to the commitment to guiding, and ultimate, values that we aim to realize. This picture of how the representation of life relates to our activities as valuing beings, of course, deeply challenges how we relate the picture of nature emerging from the sciences to ethical, aesthetic, and religious pursuits.

The concept of teleology has always hung precariously over our understanding of the course of evolution, its outcomes, and the nature of its products. The message that the question of whether or not evolution or living things are teleological, and in what sense, cannot be rationally or scientifically decided, and that it rests on our confidence in our own values will surely not be satisfying to a contemporary audience, especially one educated to value the final adjudication of objectivity by means of empirical and scientific methods. To think that our own values determine our representation of the world appears to introduce an element of subjectivity, even the threat of an element of wishful thinking, into our understanding of nature that we might view with suspicion, especially as it is an aim of science to strive for maximal and general objectivity. But when it comes to the question of the place of values and ends in life, these thinkers challenge us with a conception of human thought and agency, and their place in nature, for which the acts and capacities of subjectivity are the final testing ground and the final adjudicators. Most surprisingly to contemporary readers, perhaps, is that, for both the Life-philosophers and Rickert, the ultimacy of values and norms does not undermine our representation of a real and objective nature; it is a condition of the possibility of this representation.

To accept this picture of life and the representation of life would mean also opening the possibility that values other than knowledge *for its own sake* and procedures other than scientific analysis may also give us a window onto the real. What life *is*, what life is *after*, whether or not there is purpose or meaning in nature are questions that each sphere has a claim on because they are based on the exercise of our own diverse capacities of reasoning, intuition, and affect as we aim to realize guiding values. The difficult challenge with this picture is that the conflict among various teleological representations of evolution and of life, or between views that reject teleology altogether, is not, finally, scientifically adjudicable. If the conflict between these spheres is one that at bottom has to do with acknowledging the validity of, committing oneself to, and then ranking the guiding values of each sphere, then it is one that, even according to Rickert's rationalism, does not have a finally rational solution. Nonetheless, this is a picture of human thought and action bequeathed to us by the turn to subjectivity inaugurated by Kant's philosophy that insists, at least for "us" as knowing agents, on the primacy of value over being, the ideal over the real, and that we must still contend with today.

After Nietzschean Life-Philosophy and Rickert

The aftermath of Nietzschean biologism and Rickert's neo-Kantianism in Germany showed a variety of responses. To a large extent, the biological focus of the Life-philosophers and Rickert's rationalism were overshadowed by the rise of phenomenology on the continent. A sign of this is the extent to which the famous Davos disputation in 1929 between Martin Heidegger, the Marburg neo-Kantian Ernst Cassirer, and the logician Rudolf Carnap has come to be seen as an origin story of the major schools that have shaped twentieth- and now twenty-first-century philosophy: the Continental and Analytic traditions.[2] The focus of these recent studies shows that, in historical hindsight, the Nietzscheans and Rickert have not had the impact of these other thinkers, even though Cassirer and Heidegger took over many of the same core concerns and belonged to a common philosophical heritage. All three figures in Davos, who came to represent the major trajectories of philosophy thereafter, were far from taking Darwin, evolution, or biology very seriously, and they held that attitude for reasons similar to Rickert's. The continuity between animal and human life was thought to be irrelevant to the central philosophical problems, which were in the areas of language, logic, metaphysics, and, in Cassirer's case, symbolic form. The questions in these areas, it was thought, only pertain to and arise with human capacities of thought and forms of rationally guided agency. In Analytic philosophy, the linguistic turn and logical positivism further cemented the sense that biology was irrelevant for the core problems of philosophy.

However, even though Rickert's neo-Kantian criticism of biologism was trenchant and insightful, and is largely consistent with neo-Kantian rejections of naturalist and biological philosophy that have been carried forward in thinkers like Ernst Cassirer, Jürgen Habermas, and Karl-Otto Apel, it by no means put an end to biologism in German thought.[3] Nietzsche and his critics, even including the broader set of thinkers Rickert mentioned in his 1921 monograph, represent only a sample of the philosophers who took interest in the life sciences in Germany at the time. Indeed, they were not the only vitalist critics of mechanistic and materialist biology. For example, the non-Nietzschean vitalist Hans Driesch (1867–1941) developed what he called a philosophy of the organism that revitalized Aristotle's notion of *entelechy* to capture the *teleological* character of embryological development. For Driesch, the irreducibility of the whole organism to mechanistic interactions between its constituent parts challenged the materialist conception of life and was central to biology. In the philosophy of biology today, Driesch's call is echoed by renewed efforts to bring a teleological conception of organisms to the center of biology, and to replace the reductive and mechanistic focus on genes.

Another strand eventually resulted in the philosophical anthropology of Helmuth Plessner (1892–1985). Plessner was influenced by Husserl and another early phenomenologist, Max Scheler. Yet he explicitly combated their lack of interest in biology by attempting to ground human intentionality and consciousness in the wider biological context of organisms navigating their environments. For Plessner, human intentionality and rational agency represented a continuous outgrowth of already proto-intentional characteristics of the organic world generally. Plessner thus represented a middle way between the Nietzscheans and Rickert that insisted on the continuity between human intentional, self-reflexive agency and the wider organic world out of which these evolved. Finally, Jacob von Uexküll (1864–1944) developed an ecological philosophy based on the concept of *Umwelt* (environment) to describe the life activities of all living things. He pioneered biosemiotic and cybernetic approaches to organisms as information processors engaged in feedback loops with their environment. While all of these thinkers forged novel dialogues between biology and philosophy, these largely neglected the questions of naturalism, value, and validity, as well as the existential question of the moral meaning of nature for values, that were at the heart of the Nietzschean and Rickertian encounter with evolution.

One figure that requires special mention in the aftermath of Nietzschean Life-philosophy is Max Scheler, an important intermediary between neo-Kantianism and the emerging phenomenology of Husserl and Heidegger. While Scheler was less preoccupied with Darwin and biology, he was an avid reader of Nietzsche in his early years who went on to attend the lectures of Georg Simmel and Wilhelm Dilthey and later to develop a unique approach to value that attempted to bridge phenomenology, Life-philosophy, and Rickert's rationalism. For Scheler, all human experience is value-laden; this was simply an irreducible fact about experience that formed the foundation for the practical standpoint of seeking and reasoning about values. This phenomenological fact also formed the basis of value-driven activities of culture, including religion and science. His thought grew out of the same fin-de-siècle milieu in which biology and the sciences encountered neo-Kantian puzzles over values and normative validity. As mentioned in the introduction, Scheler wrote one of the earliest essays on Life-philosophy in 1913, and he was also responsible for bringing the work of the French vitalist Henri Bergson—a towering figure in philosophy and a defender of vitalism—to a wider philosophical audience in Germany. Scheler's phenomenological approach to value already began to downplay the relevance of biology or any of the sciences to the theory of value in a way consistent with Rickert's insistence on the unique character of human agency as driven by non-vital values. He eventually went on to challenge Nietzsche's derisive attack on Christian values of selflessness and *agape* as "anti-natural" forms of world-rejection and *ressentiment*. But he also rejected what many phenomenologists, including Husserl and Heidegger, considered the overly formal and rationalistic elements of Rickert's neo-Kantianism.

Henri Bergson has been mentioned in passing throughout this book in connection with these German discussions. He, perhaps more than anyone, was responsible for bringing vitalist criticisms of mechanistic and materialist pictures of nature to a wider audience. Bergson was hugely popular at the beginning of the twentieth century in Europe and the Unites States, more so than Nietzsche or any of his critics. He developed the notion of an *élan vital*, or vital impulse, that permeated the organic world. This creative drive was irreducible to physical interactions between the material constituents of organisms, and it was behind the movement of evolution toward new and diverse forms. Both Rickert and Scheler thought Bergson belonged in the same category as the Nietzschean Life-philosophers studied in the previous chapters. While Rickert considered Simmel to be the most sophisticated and important *Lebensphilosoph*, Bergson was more influential than any of them in defending vitalism in conversation with evolution at the turn of the century.

The impact of Edmund Husserl and then Martin Heidegger on German philosophy in the first half of the twentieth century—especially their criticisms of naturalism—for the most part eclipsed both neo-Kantianism and Life-philosophy. Even though Nietzsche and the critics surveyed here remained influential for Heidegger, this school largely replaced the vocabulary and focus with which these same problems were addressed.[4] Husserl and Heidegger were both dismissive of biological philosophy because they sought to direct philosophical attention to transcendental questions about the constitution of the world inhabited by self-conscious subjects, of which living things were only a part. They sought to direct inquiry toward a more fundamental mode of apprehension and attunement to the world that was presupposed by the positive sciences, and they downplayed continuity between human and nonhuman modes of being-in-the-world. Indeed, as we saw in the first chapter, Heidegger explicitly rejected the Darwinian, biologistic reading of Nietzsche that Overbeck and Simmel embraced. In Heidegger's terms, biology and its foundational concept of life constituted a regional ontology that was parasitic upon the fundamental ontology that he hoped to expose in his hermeneutic of human *Dasein*. Phenomenology too, then, instituted its own sharp divide between the natural sciences and philosophy, relegating Darwin and specifically evolutionary considerations to a secondary status next to the transcendental focus of philosophy. Heidegger pointed out that humans were the only beings for whom the meaning of being itself was a question, and because the question of the meaning of being was the central question of philosophy, the meaning of evolution appeared at best secondary, and at worst irrelevant.

These different trajectories with respect to the intersection between biology and philosophy emphasize the unique character of Nietzsche's Darwinism and the puzzles it raised for his critics. It was through Nietzsche that the Darwinian picture of evolution first encountered the transcendental questions about reason, agency, normativity, and value in nature that had been at the center of academic German philosophy since Kant. Many other thinkers in the German romantic and Idealist traditions—Goethe, Herder, Schelling, Hegel, Schleiermacher, Schopenhauer, Lotze, Lange, Humboldt, Trendelenburg, and Kant himself—challenged mechanistic, materialist, and reductive interpretations of the biological world. And broader debates over vitalism on the continent involving Bergson proceeded in conversation with Darwin. But the Nietzscheans were unique in confronting Darwin's picture of evolving life with the meta-ethical and metaphysical concerns over the nature and sources of value that formed the basis of sophisticated reflection on the relationship between science and religion. Indeed, the fact that Nietzsche's immoralism was grounded in assumptions about the philosophical consequences of the life sciences and evolution is a major reason for his persistence as an important interlocutor for philosophy today.

The Nietzscheans and Rickert Today

One notable bridge figure between these debates and the emergence of the philosophy of biology as a contemporary field of study was Marjorie Grene, an American philosopher who studied with Heidegger and Karl Jaspers in the early 1930s before coming to the United States for a doctorate and studying with Alfred North Whitehead. Grene is important because she largely shifted the historical orientation of what is now the contemporary philosophy of biology away from Kant and the Idealists, vitalism, and mechanism back to Aristotle. She revived Aristotle's focus on the teleological nature of organisms, but abandoned talk of this teleology as a matter of universal, vital drives. Moreover, she rejected the notion that the teleological concepts necessary for understanding organisms could be meaningfully applied to the evolutionary process.[5] In Grene's work, the field of interaction between philosophy and biology, especially on the question of teleology, thus became a confrontation between Aristotle and Darwin, and this situation of debate persists in the philosophy of biology today.

As we saw, whether or not, and in what sense, Darwinian evolution allowed for a *teleological* conception of living things became decisive for bringing it into contact with questions about normative validity and the nature of specifically human agency. Each of the thinkers in this study located teleology in nature differently, and thus each assessed its import for questions of validity in a different way. Today, the place of teleology in biology continues to be one of the central issues in the philosophy of biology.[6] Along with the problem of whether there is teleology in biology comes the question—so central to the debates recounted here—of whether defensible notions of biological teleology can form the basis of validity claims in other areas, such as ethics or epistemology. Debates in contemporary meta-ethics surrounding the possibility of natural values and natural normativity are attempting to answer many of the same questions that exercised the Nietzscheans and Rickert. And, indeed, many involved in these debates today draw heavily on Humean, Kantian, and Aristotelian thought, broadly construed, to answer them.[7] However, many differences between the basic assumptions of contemporary biologists and philosophers of biology remain a barrier for any simple translation of these turn-of-the-century debates into the contemporary context.

First, contemporary debates in the philosophy of biology about life and value operate from quite different starting points that are apparent in contemporary approaches to the basic concept of life. Most philosophy of biology textbooks dismiss the problem of the definition of life as a pseudo-problem that both the sciences and philosophers need not take seriously. Taking a pragmatic stance, these argue that lacking a definite concept of life does not prevent scientists from making progress toward understanding the biological world.[8] Moreover, contemporary accounts of teleology in biology do not take the idea of a vital drive common to all living things very seriously. Instead, teleological judgments are approached in the biological context as judgments about the "watch-like" character of organisms, the relations between their parts and the whole, and their unified functioning. Moreover, discussion of biological functions revolves around the effects of traits on survival and successful reproduction—that is, on "fitness." In this sense, these discussions reject the appetitive teleological properties of life that allowed the thinkers in this study to connect the vital-teleological context to questions about the origins of religion and science as value-driven aims. Nonetheless, biologists today often talk of living things as driven by the goal of "fitness" or reproductive success. This sort of talk is similar in important respects to Nietzsche's notion of a natural, basic drive for life, because it captures something that living things are *after*. But where biologists and philosophers of biology use this agential language, the notion that the aim of successful reproduction should at the same time be thought of as the source of life's thirst after itself is viewed as mere metaphor and rhetorical license. The Darwinian *telos* of fitness is not that of the biologized *eros* that Nietzsche found at the origins of cultural values, but the fact that living processes are conceived as fundamentally after something continues to be important for relating evolution to ethical and religious life.[9]

These largely deflationary approaches to the concept of life and vital-teleology undermine the particular role these concepts might play in justifying the idea of natural values and of life itself as a natural value, except to remark that most living things behave *as if* they want to survive and reproduce. This would be far too little to justify Nietzsche's notion of life-affirmation as a criterion of health that can be used as the basis for a revaluation of values. For this reason, one might think that contemporary philosophers would be more in favor of the Rickertian line, in which *validity* can be meaningfully spoken of solely within the province of rational, self-legislative, agency that is responsive to reasons. The standard neo-Darwinian picture of organisms driven toward reproductive success salvages a minimal notion of teleology, but this is clearly not one that can easily be understood as the source of intrinsic, non-instrumental values like truth *for its own sake* or the existential aim of a way of life in harmony with reality. Thus, there is still room to investigate whether and how the intrinsic validity of values that appear autonomous from mere biological life can be vindicated with reference to the Darwinian *telos* of reproductive success. These questions and problems have resurfaced in Thomas Nagel's recent discussions of the problems with materialist views of evolution and life with respect to consciousness, epistemic and ethical value, and knowledge. Nagel argues that the insufficiency of current physicalist theories for understanding mind, value, and knowledge as non-accidental features of nature require that we take seriously the teleological view of the universe as biased toward the generation of life.[10]

In addition to these contemporary resonances, neo-Aristotelian virtue ethicists such as Michael Thompson, Philippa Foot, and Rosalind Hursthouse have revitalized the notion of natural value by grounding it, as Nietzsche does, in the teleological character of living things.[11] Foot's concept of "natural goodness" aims to recover the sense that biological and ethical judgments share the same logical basis and structure, and that the teleological character of organisms justifies notions of well-being. For Thompson and Foot, judgments about ethical virtues and failings are of the same kind, and they have the same basis, as judgments of natural function and dysfunction of vital processes. However, Foot recognized that Nietzsche presents an interesting adversary because, while he accepted the notion of natural values grounded in biological teleology, he rejected the view that these natural values or virtues are those of altruistic morality.[12] The views of natural teleology in modern Aristotelian writers are not without their forceful critics, who, like Rickert, deny any place at all for ethical or even proto-ethical values in the nonhuman world and in nature.[13] Yet these current debates show that the problems at the basis of Nietzsche's response to Darwin continue to be lively areas of philosophical debate that touch upon a wide range of problems in meta-ethics, normative ethics, and evolutionary theory. While our scientific understanding of the mechanisms and processes of evolution has changed significantly since the turn of the twentieth century, the philosophical issues it raises have not. These continue to be crucial topics for future investigation and debate.

Philosophy, Science, and Religion

This study has shown that debates over life and value that responded to Darwin naturally bled into discussion of the origins of concrete ideals in cultural history. And these thinkers focused in particular on the natural-historical origins of values that they argued were constitutive both of many concrete religious practices and beliefs and of the core values embedded in scientific investigation. This aspect of the debates between Nietzsche and his critics differs in crucial respects from contemporary debates over science and religion that span the cognitive, evolutionary, anthropological, and social-scientific studies of religion today. And this is because these debates focused on a particularly philosophical element of religions, and a religious element of philosophy that coheres around the existential question of the meaning of nature that I have used to frame this investigation. I would like to highlight three points that might be taken away from these discussions to broaden the perspective of contemporary reflection on science and religion.

The first lesson again arises out of the general rejection of a positivist fact-value distinction that is a remnant of Weber and nineteenth-century positivism and that is still often used to demarcate value-free, scientific, and objective science from nonscientific work.[14] In the positivist conception, scientific and rational judgment ends where one begins to make evaluative claims or claims about what is rational, rather than purely descriptive, causal, or predictive ones. Overbeck was closest to accepting a positivist distinction like this, but he too argued (as Weber later did) that the purely historical perspective was an exercise in realizing the ideal of freedom. His distinction between science and life had to do with the difference between purely theoretical interests and existential or vital interests in the meaningfulness and value-laden character of life, not between facts and values. The Life-philosophers and Rickert provide a powerful philosophical perspective because they started from the position that all human thought and action was *teleological* and, therefore, *value-laden*. For this reason, even purely theoretical activities that aimed at value-neutral description and causal explanation implicitly posited values. The key point was that the value-laden character of knowledge did not invalidate the sciences; it made them possible. The sciences could not eschew or avoid meta-ethical and metaphysical investigation of the sources and validity of values, or into the meaningfulness of human activities, because they too presupposed this.

For these thinkers, science was inquiry into truth *for its own sake*. It was a practice that cared for, in Rickert's sense of culture as *Pflege*, this noninstrumental value. Science thus belonged within human ethical life as a practice that extended and cultivated certain human capacities at the same time that it generated something of value *in* and *for* itself—namely, understanding of nature. Scientific knowledge was value-free only in the sense that the measure of reality was not to be assessed according to extra-scientific moral, aesthetic, or religious aims. Thus, value-freedom from non-epistemic values was itself an ideal and a value, not a neutral starting point free from subjectivity or the strivings of living agents.

Similarly, all of these thinkers answered the question of what religion is or, perhaps less generally, of what the core concepts of various particular religions were by trying to identify the constitutive values at which various religions aimed. Instead of defining religions exclusively in terms of belief in supernatural agents that explained natural occurrences, these thinkers focused on seeking values of purity, holiness, perfection, and the value of life itself. They sought a representation of life that could itself play a role in understanding and realizing these values in nature.[15] Indeed, these thinkers claimed that values and metaphysical views about the nature of reality as a whole were intimately connected because metaphysical views about reality could not be formulated independently of stances toward the reality of value. Religions were examples of a form of transcendental reasoning that proceeded from experienced values and commitments to their validity to claims about the way the world must be for the validity of these values to be possible and justified. The source of religions, in the terms of the Nietzscheans, was life's own attempt to comprehend the basis of its own valuing of itself. In Rickert's terms, religions sprang from reason's search for a completion in all arenas of human life and for a conception of the world in which this completion was intelligible. In the metaphysical and pre-Kantian views of Platonists and Aristotelians, values could be conceived to simply belong to reality, whether as inseparable parts of the natural world or as a transcendent order. While many contemporary thinkers have turned to pragmatism as a way of avoiding both metaphysics and the forms of transcendental thinking on display here, it is important to see this question of the relationship between value and reality as a question at the center of contemporary philosophical divisions.[16]

These views about religion and science, and the problems they raise, construe the alleged "conflict" between them in a particular manner. Science and religion could conflict if the sciences could offer a positive or negative answer to the existential question of the meaning of reality *for us*. By arguing that science pursues a different kind of conception of reality that answers to theoretical interests rather than practical or existential interests, these thinkers argued that this question could neither be intelligibly formulated nor answered within the sciences. From the scientific perspective, this existential question appears meaningless or ill-formulated. The picture of nature generated by the sciences is objective insofar as it is divorced from questions of meaningfulness, validity, and value. To seek to answer such a question, the Life-philosophers and Rickert show us, the positive sciences would stray into meta-ethical and metaphysical questions about the nature and reality of value.[17] They would have to raise their own fundamental, normative commitments to awareness and critique, and in so doing leave behind empirical investigation and theoretical explanation.

All of these thinkers also endorsed the point that this value-laden character of inquiry did not undermine validity claims; rather, it enabled them and constituted the internal rationality of each sphere of human activity with its unique guiding values. Just as Kant's critiques taught that there is an irreducibly normative element to all conceptual understanding, practical reasoning, and even (more controversially) aesthetic and biological judgment, so the Life-philosophers and Rickert extended this insight to culture as a whole. In Rickert's terms, which could apply to all of the Life-philosophers as well, "culture" was synonymous with care (*Pflege*) for ideals and values. Challenging a positivist view of the distinction between facts and values, science and nonscience, rationality and irrationality, is the crucial first step in recasting the conflict between science and religion to bring out the fundamentally philosophical issues that emerge when science itself is seen to be a value-laden ideal of life.

This has implications for contemporary areas of the sciences, especially evolution biology, cognitive science, and neuroscience, which aim at a scientific explanations of religion.[18] Of course, I am not arguing here that that a naturalistic account of values is necessarily flawed. Indeed, naturalistic and evolutionary accounts of various spheres of value today are much more nuanced than the accounts that Nietzsche, Overbeck, and Simmel provided. But I am arguing that the answers to the meta-ethical and metaphysical questions about agency, the nature of life, teleology, normativity, and the nature of rationality that these topics raised are far from obvious, and they require us to ask questions that go beyond what the sciences tell us. There are many live philosophical options for dealing with these issues, but the procedure for choosing among them is much less clear, and there is room for reasonable disagreement. Moreover, positions on these matters are often presupposed when scientists offer explanations of cultural phenomena like religions, ethical values, and even science itself. The task of scientifically investigating and understanding religion, or even scientifically explaining science itself, inevitably leads to the philosophical problems of value and life at the heart of the debates recounted in this book.

A few recent works defend more philosophically sophisticated approaches to the relationship between science and religion that have been outlined here, but these often go unacknowledged against the overpowering and less philosophically sophisticated voices of many who write about science and religion today.[19] This is especially true of the popular writings on these topics. The discussions of science and religion that followed in the wake of the so-called New Atheists —Sam Harris, Richard Dawkins, Lawrence Krauss, and Christopher Hitchens—suffer from a lack of philosophical sophistication. These popular debates raise issues in the philosophy of science, the philosophy of religion, and the theory of value without the conceptual tools or nuance necessary to answer them adequately. This is likely because giving a more sophisticated view of the different sorts of questions asked and answered in scientific investigation, in philosophical analysis, and in religious practice would force us to take the meta-ethical and metaphysical investigation more seriously and carefully than these thinkers do. Appreciating the complexity of the debates about religion and science that occurred at the end of the nineteenth century can help us see the deep problems that lie in any attempt to sort out the relationship between philosophy, science, and religious thought. And it can prevent us, as so often happens, from formulating this relationship in crude terms of "faith" versus "fact."[20] These thinkers show that in bringing science into contact with religion, both can and must change in the process.

In raising philosophical questions about forms of agency and value and their place in nature, these thinkers brought the life sciences into conversation with the fundamental questions that have shaped Western philosophy and religious thought. I have suggested that this focus on agency and value invited scientific approaches to religion to critically examine fundamental assumptions about what religions are. In doing so, it also brought purely scientific interests into contact with extra-scientific interest in questions of meaning and value. Because the question of human agency and value is so close to us that it is difficult to maintain the disinterested distance required for inquiry into the truth *for its own sake*, it is one that makes this division between scientific and extra-scientific interest increasingly difficult to maintain. This is an area in which our own fundamental commitments to values are always in play and must be in play. It is an area in which solutions are supplied only by the evidence of what satisfactions humans seek, what ideals they commit themselves to, and what they are indeed able to achieve in any area of life. These issues thus raise questions that go well beyond narrow disciplinary boundaries and the specialized topics that occupy most working scientists, but they are nonetheless fundamental for understanding human life, its capacities, and its possibilities. As we contiune to probe human agency and understanding and their place in the order of nature in conversation with our own normative commitments to values, our scientific pursuits will be difficult to separate from our confidence in the validity of those values that guide our extrascientific lives.

Acknowledgments

There are too many to thank for the intellectual comradery, critical insight, and support that have made this book possible. First, I would like to thank the Deutsche Akademische Austausch Dienst (DAAD) for a generous fellowship that allowed me to conduct research in Germany, where I benefited enormously from seminars, conferences, and conversations at Albert-Ludwigs-Universität, Freiburg. I would especially like to express my gratitude to Andreas Urs Sommer for his support, hospitality, and critical exchanges, and for commenting on early drafts of chapters. I am grateful that I had the opportunity through many engaging conversations to learn from him, and especially his work on Nietzsche and Overbeck. I would also like to thank Virginie Palette, Jethro Masis, and Lucian Lionel for always lively and insightful dialogue about philosophy—and everything besides.

In the last years of writing and refining the arguments of the book, I have benefited in many ways from the intellectual community at the University of Cambridge. I am especially grateful for the support of the Templeton World Charity Foundation and of Sarah Coakley, whose ingenious and unconventional idea to place scholars of religion into active research groups in the sciences gave me the opportunity to develop the ideas in this book in conversation with her and with leading evolutionary biologists. I would like to thank Daniel De Haan and Natalja Deng for helpful dialogue, Tim Clutton-Brock for being open to this project and hosting me in his lab, and Hasok Chang for support and for integrating me into the Department of History and Philosophy of Science. I would also like to thank all of the members of the Large Animal Research Group with whom I have enjoyed talking philosophy, science, and religion over these last years, in particular Dieter Lukas and Alecia Carter.

I cannot possibly account for all that I learned from my mentors and colleagues at Stanford, in particular Tom Sheehan, Shahzad Bashir, Hester Gelber, Nadeem Hussain, and Noreen Khawaja. They were all incredibly helpful in the development of this project. I would like to thank in particular Lee Yearley for his knowledge, wise guidance, and provocative questioning during various phases of my writing and research and for commenting on early drafts.

The person who deserves the most thanks for support, mentorship, and conversations along the way is Brent Sockness. It is hard to overestimate how much this project has benefited from Brent's encouragement and from our many discussions. Brent's support and his comments on various drafts helped me refine and sharpen the text considerably. I take full responsibility for any shortcomings that remain despite Brent's mentorship and the help of all of the generous and brilliant people with whom I have been so fortunate to work through the years of writing this book.

Finally, I would like to express special thanks to my family. It is extraordinarily difficult to find the right words to express how much their support over the years has meant to me. Thanks to my aunt, Gerda Wagner, for giving me a welcome home and for many enjoyable days and nights of enlightening conversation. Thanks also to my cousins Ingo, Uwe, and Jürgen for support during my time in Germany. Thanks to my brother Greg and his wife, Samantha, for always being there. Last, it is not possible to do justice to the humor, patience, and wisdom of my parents, Peter and Helga, to whom I owe so much. I dedicate this book to them.

Notes

Introduction

1. For a fascinating history of the discovery of deep time both before and after Darwin, see Martin Rudwick, *Earth's Deep History: How It Was Discovered and Why It Matters* (Chicago: University of Chicago Press, 2014).

2. For a classic argument that Darwin's lesson for philosophy and religion is the absence of intelligent design and purpose in nature, see Daniel Dennett's now classic *Darwin's Dangerous Idea: Evolution and the Meanings of Life* (New York: Simon & Schuster, 1996). For a more recent argument that one of the most significant implications of Darwin's theory was its ability to account for design without invoking a designer, see Philip Kitcher, *Living with Darwin: Evolution, Design, and the Future of Faith* (Oxford: Oxford University Press, 2009).

3. Immanual Kant, *Critique of the Power of Judgment*, ed. Paul Guyer, trans. Paul Guyer and Eric Matthews (Cambridge: Cambridge University Press, 2002), 271/5:400. Citations to the English translations of Kant's texts are in the following format: [page number] / [section number]:[page number of original critical edition]. The question of whether a mechanistic conception of organisms is possible is still a crucial topic of debate in the philosophy of biology.

4. For excellent overviews of the many facets of the reception of Darwin and Darwinism across Europe, see Thomas F. Glick, ed., *The Comparative Reception of Darwinism* (Chicago: University of Chicago Press, 1988); for debates in the English context, see John Hedley Brook, "Darwin and Victorian Christianity," in *The Cambridge Companion to Darwin* (Cambridge: Cambridge University Press, 2009), 197–218.

5. For a highly informative study of German romantic *Naturphilosophie* and its influence on Darwin, see Robert Richards, *The Romantic Conception of Life: Science and Philosophy in the Age of Goethe* (Chicago: University of Chicago Press, 2002). The tradition Richards traces in German philosophy of nature provides informative historical and cultural background to Nietzschean Life-philosophy.

6. Recent debates in philosophy have addressed the implications of Darwinism for the theory of value and meta-ethics. See Sharon Street, "A Darwinian Dilemma for Realist Theories of Value," in *Philosophical Studies* 127, no. 1 (2006): 109–66. Street argues that evolutionary theory poses insuperable problems for metaphysically realist theories of value that see values as objective, normative facts. See Thomas Nagel's reply in *Mind and Cosmos* (Oxford: Oxford University Press, 2012), 97–125. Nagel's riposte is that if (neo-)Darwinian theory undermines the reality

and objectivity of epistemic and moral value, then this theory must be getting something wrong. Many of the basic concerns of Nietzsche and his critics about the ethical implications of Darwin, and about the relationship between moral and scientific objectivity, resurface in contemporary exchanges over evolutionary "debunking" arguments.

7. Rickert uses this polemical language in his book on Life-philosophy, which offers a neo-Kantian critique of the philosophical foundations of the theory of value in Nietzsche, Overbeck, and Simmel.

8. Peter Kropotkin, *Mutual Aid: A Factor of Evolution*, ed. Paul Avrich (New York: New York University Press, 1972).

9. For an example of an influential approach to the evolution of cooperation that does not explicitly engage with meta-ethical or broadly metaphysical issues, see Martin Nowak, "Five Rules for the Evolution of Cooperation," *Science* 314 (2006): 1560–63. Nowak's game-theoretic approach has recently caused controversy because it rejects the mathematical and conceptual basis of Hamilton's Rule, which had been the reigning theoretical approach to the evolution of cooperation and altruism to date. See Martin Nowak, Corina Tarnita, and E. O. Wilson, "The Evolution of Eusociality," *Nature* 466 (2010): 1057–62, and the critical reply signed by 137 evolutionary biologists that followed it in Patrick Abbot, Jun Abe, John Alcock, et al., "Inclusive Fitness Theory and Eusociality," *Nature* 471 (March 2011): E9–E10.

10. Thomas Scott-Phillips, Thomas Dickins, and Stuart West, "Evolutionary Theory and the Ultimate–Proximate Distinction in the Human Behavioral Sciences," *Perspectives on Psychological Science* 6, no.1 (2011): 38–47.

11. There is a large amount of recent literature on cooperation and its importance for the evolution of morality and religion in anthropology, evolutionary biology, philosophy, and the study of religion. For a few recent and representative examples of this work, see Ara Norenzayan, *Big Gods: How Religion Transformed Cooperation and Conflict* (Princeton, NJ: Princeton University Press, 2013); Ara Norenzayan, Joseph Henrich, and Edward Slingerland, "Religious Prosociality: A Synthesis," in *Cultural Evolution: Society, Technology, Language, and Religion*, ed. Peter J. Richerson and Morten H. Christiansen (Cambridge, MA: MIT Press, 2013); and Bernard Crespi and Kyle Summers, "Inclusive Fitness Theory for the Evolution of Religion," in *Animal Behavior* 92 (June 2014): 313–23.

12. For contemporary accounts of this issue, see Helen Longino's *Science as Social Knowledge: Values and Objectivity in Scientific Inquiry* (Princeton, NJ: Princeton University Press, 1990) and the essays in *Value-Free Science?: Ideals and Illusions*, ed. Harold Kincaid, John Dupre, and Alison Wyle (Oxford: Oxford University Press, 2007).

13. See, for example, the recent book of this genre by Jerry Coyne, *Faith vs. Fact: Why Science and Religion Are Incompatible* (New York: Penguin Press, 2015).

14. This term relates to Stephen Jay Gould's now classic doctrine that science and religion are "non-overlapping magisteria," or NOMA, so that each has its own domain of legitimate teaching authority. See Stephen Jay Gould, "Non-Overlapping Magisteria," *Natural History* 106 (March 1997): 16–22. Nietzsche and his critics would challenge Gould's strict separation of "facts" from "values" or "meanings"; how they do so is analyzed in each of the following chapters.

15. Although this urgency appears in many thinkers, both before and after World War I, this feeling of a cultural crisis is given representative expression in Ernst Troeltsch's 1922 essay "The Crisis of Historicism." See Ernst Troeltsch, "Die Krisis des Historismus," in *Ernst Troeltsch Kritische Gesamtausgabe*, vol. 15, ed. Gangolf Hübinger (Berlin: De Gruyter, 2002), 433–56.

16. This narrative is not intended to suggest that Hegel's philosophy was, or has been, refuted. Hegel's conception of self-reflexive rationality is defended by a number of highly influ-

ential philosophers as a live philosophical option for understanding the relation between mind, normativity, and nature. See, for example, Robert Pippin, *Idealism as Modernism: Hegelian Variations* (Cambridge: Cambridge University Press, 1997), and *Hegel's Idealism: The Satisfactions of Self-Consciousness* (Cambridge: Cambridge University Press, 1989); Terry Pinkard, *Hegel's Naturalism: Mind, Nature, and the Final Ends of Life* (Oxford: Oxford University Press, 2012); Robert Brandom, *Tales of the Mighty Dead: Historical Essays on the Metaphysics of Intentionality* (Cambridge, MA: Harvard University Press, 2002).

17. See Heinrich Rickert, *The Limits of Concept-Formation in Natural Science*, ed. and trans. Guy Oakes (Cambridge: Cambridge University Press, 1986).

18. See Frederick Beiser, *The German Historicist Tradition* (Oxford: Oxford University Press, 2011) for a highly informative study of the key problems of German philosophy during the long nineteenth century.

19. Friedrich Nietzsche, *Beyond Good and Evil*, ed. and trans. Marion Faber (Oxford: Oxford University Press, 1998), 24.

20. See Charles Taylor, *A Secular Age* (Cambridge, MA: Harvard University Press, 2007) for a fascinating history of the rise of secular culture in the West. The history of "secularization" and the problems with various secularization narratives are still the subjects of debate and investigation among scholars of religion.

21. See J. Samuel Preus, *Explaining Religion: Criticism and Theory from Bodin to Freud* (Oxford: Oxford University Press, 1996).

22. For a fascinating history of debates over the status of theology with regard to science in Germany during this period, see Johannes Zachhuber, *Theology as Science in Nineteenth-Century Germany: From F. C. Baur to Ernst Troeltsch* (Oxford: Oxford University Press, 2013).

23. Max Scheler, "Versuch einer Philosophie des Lebens," in *Umsturz der Werte*, 4th ed. rev., vol. 3 of *Gesammelte Werke* (Bern: Francke, 1955), 311–41.

24. This tendency to see a close relationship between Life-philosophy and pragmatism existed very early and persists in contemporary thought. Ernst Troeltsch wrote of the close "affinity" between William James and Henri Bergson in his 1912 essay "Empiricism and Platonism in the Philosophy of Religion" (*Harvard Theological Review* 5, no. 4 [1912]: 401–23). Hans Joas also interprets Life-philosophy as the German analogue of American pragmatism for the priority that both accord to practice over theory. Joas, however, interprets *life* to mean *creativity* and so downplays the evolutionary resonance of this notion in the life-sciences and in the project of naturalism. See Hans Joas, *The Creativity of Action* (Chicago: University of Chicago Press, 1996), 116–26.

25. See Jürgen Große, *Lebensphilosophie* (Stuttgart: Reclam Verlag, 2010).

26. See Robert Richards, *The Romantic Conception of Life* (Chicago: University of Chicago Press, 2002).

27. Große, *Lebensphilosophie*, 23.

28. Jacobi is an important and fascinating figure who wrote at the same time as Kant was publishing his *Critiques*. Jacobi's unique Christian appropriation of Hume first announced the problem of nihilism that was later to become central for Nietzsche. He developed his position in opposition to a rationalist Spinozism during the famous "Pantheism controversy" (*Pantheismusstreit*) over the alleged pantheist views of the famous German literary figure, Gotthold Emphraim Lessing.

29. This narrative leaves out important figures who were deeply involved in these problems, but whose views are difficult to place in this scheme. I leave out Schelling because his synthesis of naturalism and idealism was an important precursor to the turn to philosophy of nature

(*Naturphilosophie*) in Life-philosophy that is difficult to situate. I also leave out Schleiermacher, whose turn to "feeling" and the distinctively religious feeling of absolute dependence is also a rejection of the autonomy and spontaneity of reason characterized by Kant and Hegel. Andrew Dole argues that Schleiermacher was a religious naturalist, but this issue remains controversial due to the tension between Schleiermacher's naturalism and his Idealist and Kantian commitments. See Andrew Dole, *Schleiermacher on Religion and the Natural Order* (Oxford: Oxford University Press, 2010).

30. Recent debates in biology that have controversially revived the ghost of Lamarck and the inheritance of acquired traits have focused on the role of epigenetic inheritance. See Eva Jablonka and Marion Lamb, *Evolution in Four Dimensions: Genetic, Epigenetic, Behavioral, and Symbolic Variation in the History of Life* (Cambridge, MA: MIT Press, 2005).

31. Hans Joas, *The Genesis of Values*, translated by Gregory Moore (Chicago: University of Chicago Press, 2000), 21.

32. See Herbert Schnädelbach, *Philosophy in Germany, 1831–1933*, trans. Eric Matthews (Cambridge: Cambridge University Press, 1984), 198–99. Lotze is a fascinating figure in his own right, and his student, Wilhelm Windelband, who was Rickert's teacher, initiated the development of Southwest, or Baden, School of neo-Kantianism.

33. Lotze was an early critic of the notion of "vital forces" in the organic world, but he defended a teleological notion of organisms through the Kantian principle that teleology is a regulative idea for guiding our inquiry into nature, not a constitutive principle of nature itself. See Frederick Beiser, *Late German Idealism: Trendelenburg and Lotze* (Oxford: Oxford University Press, 2013), 261–83.

34. Nietzsche's relation to Kant and the post-Kantian tradition is complex and has been explored by a number of scholars who have argued that early neo-Kantians like Friedrich Albert Lange and African Spir were especially influential for him. See Tom Bailey, "Nietzsche the Kantian?" in *The Oxford Handbook of Nietzsche*, ed. Ken Gemes and John Richardson (Oxford: Oxford University Press, 2013), 134–59; Kevin R. Hill *Nietzsche's Critiques: The Kantian Foundations of His Thought* (Oxford: Oxford University Press, 2003). For the significant influence of the Kantian Friedrich Albert Lange on Nietzsche, see Nadeem Hussain, "Nietzsche's Positivism," in *European Journal of Philosophy* 12, no. 3 (2004): 326–68; and G. J. Stack, "Kant, Lange, and Nietzsche: Critique of Knowledge," in *Nietzsche and Modern German Thought*, ed. Keith Ansell Pearson (New York: Routledge, 1991), 30–58.

35. Heinrich Rickert, *System der Philosophie: Allgemeine Grundlegung der Philosophie* (Tübingen: Mohr-Siebeck, 1921), 153.

36. Thomas Nagel, *Secular Philosophy and the Religious Temperament: Essays 2000–2008* (Oxford: Oxford University Press, 2010), 5.

37. Schubert Ogden, *On Theology* (Dallas, TX: Southern Methodist University Press, 1986), 122.

38. Kevin Schilbrack, *Philosophy and the Study of Religions: A Manifesto* (Cambridge, MA: Wiley-Blackwell, 2014), 135.

39. This is perhaps where the Nietzschean and Rickertian lines of argument depart most explicitly from contemporary evolutionary explanations of religion. Using inclusive fitness, kin-selection, and evolutionary game theory, recent work in the sciences defines religion in terms of reward and punishment mechanisms, and it approaches the evolutionary explanation of religion by showing its so-called ultimate role in promoting behavior that has adaptive utility for survival and reproduction. See Norenzayan, *Big Gods*, 4–8.

40. Kant, *Critique of the Power of Judgment*, 233/5:360.

41. See, for example, a recent compilation of essays debating teleology in biology: Andre Ariew, Robert Cummins, and Mark Perlman, eds. *Functions: New Essays in the Philosophy of Psychology and Biology* (Oxford: Oxford University Press, 2002). For an important and sophisticated contemporary defense of a teleological view of organisms, see Denis Walsh, *Organisms, Agency, and Evolution* (Cambridge: Cambridge University Press, 2015).

42. Friedrich Nietzsche, *Zur Genealogie der Moral*, in *Kritische Studienausgabe*, vol. 5, edited by Giorgio Colli and Mazzino Montinari (Berlin: De Gruyter, 1988), 313–16.

43. Robert Cummins, for example, rejects this notion in his essay "Neo-Teleology" in Ariew, Cummins, and Perlman, *Functions*, 157–72.

44. Friedrich Nietzsche, *Jenseits von Gut und Böse*, in *Kritische Studienausgabe*, vol. 5, ed. Giorgio Colli and Mazzino Montinari (Berlin: De Gruyter, 1988), 169; Franz Overbeck, *Ueber die Christlichkeit unserer heutigen Theologie*, in *Franz Overbeck Werke und Nachlass*, vol. 1, ed. Ekkehard W. Stegemann and Niklaus Peter (Stuttgart: J. B. Metzler, 1994); Georg Simmel, *Lebensanschauung* (München: Duncker & Humboldt, 1918), 26.

Chapter One

1. Nietzsche, *Die Geburt der Tragödie*, in *Kritische Studienausgabe*, vol. 1, ed. Giorgio Colli and Mazzino Montinari (Berlin: De Gruyter, 1988), 19. The following citations in this chapter continue to reference the version of Nietzsche's texts found in his *Sämtliche Werke: Kritische Studienausgabe in 15 Bänden*, ed. Giorgio Colli and Mazzino Montinari, 15 vols. (Berlin: De Gruyter, 1988). I cite these volumes using the title of the work and the abbreviation *KSA*, followed by the volume and page reference (i.e., *KSA*, 1:19). All translations are mine unless otherwise specified.

2. Bernard Williams understood Nietzsche's literary strategy to be an intentional dismissal of traditional ways of doing philosophy, which was justified by his understanding of psychology. See Bernard Williams, "Nietzsche's Minimalist Moral Psychology," in *Nietzsche, Genealogy, Morality: Essays on Nietzsche's "On the Genealogy of Morality,"* ed. Richard Schacht (Berkeley: University of California Press, 1994), 238.

3. Christopher Janaway convincingly argues that Nietzsche's rhetorical and literary style is tied to his theory of the role of affect in judgment. Nietzsche's aim is to educate and excite the affects of his readers, to win them over through appeal to their affects rather than their reason; see "Naturalism and Genealogy," in *A Companion to Nietzsche*, edited by Keith Ansell Pearson (Oxford: Blackwell Publishing, 2006), 337–52.

4. See Andreas Sommer, "Ein Philosophisch-Historicher Kommentar zu Nietzsches *Götzen-Dämmerung*: Probleme und Perspektiven," in *Perspektiven der Philosophie* 35 (2009): 45–66.

5. For an example of this claim, see Bruce Knauft, *Genealogies for the Present in Cultural Anthropology* (New York: Routledge, 1996), 84–88. Knauft writes that "the logical conclusion of cultural studies—at least its emergent American form—is the work of Friedrich Nietzsche." See also the Foucauldian form of genealogical critique and analysis of religion practiced by Talal Asad in *Genealogies of Religion: Discipline and Reasons of Power in Christianity and Islam* (Baltimore: Johns Hopkins University Press, 1993); and also *Formations of the Secular: Christianity, Islam, Modernity* (Stanford, CA: Stanford University Press, 2003).

6. See especially the original challenge to the postmodern Nietzsche in Maudemarie Clark, *Nietzsche on Truth and Philosophy* (Cambridge: Cambridge University Press, 1990). For studies on Nietzsche and the life sciences that have informed this chapter, see Gregory Moore, *Nietzsche, Biology, and Metaphor* (Cambridge: Cambridge University Press, 2002). John Richardson's *Nietzsche's New Darwinism* (Oxford: Oxford University Press, 2004) showed the importance of

Darwinian themes for Nietzsche's project. Another recent study is Christian Emden's *Nietzsche's Naturalism: Philosophy and the Life-Sciences in the Nineteenth Century* (Cambridge: Cambridge University Press, 2014).

7. The naturalistic reading of Nietzsche has surfaced in recent Anglo-American scholarship in the writings of Brian Leiter, John Richardson, Maudemarie Clark, Peter Poellner, and Richard Schacht, among others. This presentation of Nietzsche's mediation of religion and science has been especially informed by recent work in philosophy on the concept of life and on Nietzsche's Darwinism. See especially Nadeem Hussain, "The Role of Life in the *Genealogy*," in *The Cambridge Critical Guide to Nietzsche's "On the Genealogy of Morality*," ed. Simon May (Cambridge: Cambridge University Press, 2011), 142–69; and John Richardson, "Nietzsche on Life's Ends," in *The Oxford Handbook of Nietzsche*, ed. Ken Gemes and John Richardson (Oxford: Oxford University Press, 2013), 756–84. For a rigorous and insightful analysis of Nietzsche's metaphysics against the background of neo-Kantian thought, see Peter Poellner, *Nietzsche and Metaphysics* (Oxford: Clarendon Press, 1995).

8. See Moore, *Nietzsche, Biology, and Metaphor*, 21–55.

9. Gregory Moore shows Nietzsche's language of the vitality and degeneracy of nerves, drives, and instincts to be part of a much broader "medicalization" and "biologization" of philosophical discourse in the nineteenth century. See Moore, *Nietzsche, Biology, and Metaphor*, 115–28.

10. Friedrich Nietzsche, *Unzeitgemässe Betrachtungen*, KSA, 1:269.

11. Historicism presented just as important a problem for ethical and religious thought in this generation of thinkers as did evolutionary naturalism. For a thorough study of the problem of historicism in German thought, see Beiser, *German Historicist Tradition* (Oxford: Oxford University Press, 2011).

12. Nietzsche, *Unzeitgemässe Betrachtungen*, KSA, 1:296.

13. Nietzsche, *Die Geburt der Tragödie*, KSA, 1:12–13.

14. Nietzsche, *Die Geburt der Tragödie*, KSA, 1:14.

15. See Andreas Sommer, *Der Geist der Historie und das Ende des Christentums: Zur "Waffengenossenschaft" von Franz Overbeck und Friedrich Nietzsche* (Berlin: Akademie Verlag, 1997), 17–28, for an informative discussion of how Nietzsche's project of recovery related to the disciplinary norms of academic philology.

16. Nietzsche, *Ecce Homo*, KSA, 6:310.

17. Burckhardt argued in his book *History of Greek Culture* against the overly scientific conception of history promoted by contemporary scholars. He wrote that cultural history "has as its object the inner life of past humanity and it describes how that humanity existed, desired, and thought, how it looked upon the world and how it was able to act on it." See Jacob Burckhardt, *History of Greek Culture*, trans. Palmer Hilty (New York: Dover Press, 2002), 330.

18. Nietzsche, *Ecce Homo*, KSA, 6:307.

19. Ibid.

20. Ibid., 6:312.

21. Ibid.

22. Ibid., 6:311.

23. Ibid., 6:310.

24. Ibid., 6:311.

25. Nietzsche, *Götzen-Dämmerung*, KSA, 6:120.

26. Andreas Sommer, "Nietzsche und Darwin," in *Nietzsche als Philosoph der Moderne*, ed. Barbara Neymeyer and Andreas Sommer (Heidelberg: Universitätsverlag, 2012), 223–40.

27. For a list of the books on biology that he owned and read carefully, see the catalogue of Nietzsche's personal library in Guiliano Campioni, Paolo D'Iorio, Maria Cristina Fornari, et al., eds., *Nietzsches Persönliche Bibliothek* (Berlin: Walter de Gruyter, 2003).

28. Heinrich Rickert, *Die Philosophie des Lebens: Darstellung und Kritik der Modeströmungen unserer Zeit* (Tübingen: J. C. B. Mohr, 1920), 97.

29. See, for example, David Marc Hoffmann, *Zur Geschichte des Nietzsche-Archivs* (Berlin: de Gruyter, 1991), 479; and Moore, *Nietzsche, Biology and Metaphor*, 46–55.

30. Andreas Sommer, "Nietzsche mit und gegen Darwin in den Schriften von 1888," in *Nietzsche, Darwin, und die Kritik der Politischen Theologie*, ed. Volker Gerhardt and Renate Reschke (Berlin: Akademie Verlag, 2010), 44.

31. There is evidence that Nietzsche entertained the idea that the teleological drive of "life" and a "will to power" was not limited to the organic world and applied to all of physical reality. See Moore, *Nietzsche, Biology and Metaphor*, 42–43.

32. Ibid., 37–40.

33. Ibid., 29.

34. Nietzsche, *Jenseits von Gut und Böse*, KSA, 5:27.

35. Nadeem Hussain argues convincingly that this argument is invalid because it does not answer the normative question still left open of why life "ought" to be valued, especially if it is driven toward non-moral ends. That only life can genuinely satisfy a basic drive and "hunger" still does entail that it "ought" to be affirmed. See Hussain, "The Role of Life in the *Genealogy*," 162. To foreshadow later chapters, a similar objection to the way Nietzsche's concept of life straddles the "is" and the "ought" resurfaces in Simmel and Rickert.

36. Nietzsche, *Genealogie*, KSA, 5:411.

37. Nietzsche, *Götzen-Dämmerung*, KSA, 6:71–72.

38. Nietzsche, *Die Fröhliche Wissenschaft*, KSA, 3:340.

39. Nietzsche, *Götzen-Dämmerung*, KSA, 6:68

40. For an example of some of contemporary debates over Nietzsche's meta-ethics, see Simon Robertson, "Normativity for Nietzschean Free-Spirits," in *Inquiry* 54, no. 6 (2011): 591–613; Brian Leiter, "Nietzsche's Metaethics: Against the Privilege Readings," *European Journal of Philosophy*, 8, no. 3 (2000): 277–97; Nadeem Hussain, "Honest Illusion: Valuing for Nietzsche's Free Spirits" in *Nietzsche and Morality*, ed. Brian Leiter and Neil Sinhababu, (Oxford: Clarendon Press, 2007), 157–91; and Maudemarie Clark and David Dudrick, "Nietzsche and Moral Objectivity: The Development of Nietzsche's Metaethics," in *Nietzsche and Morality*, ed. Brian Leiter and Niel Sinhababu (Oxford: Clarendon Press, 2007), 192–226; for a volume dedicated to this discussion, see Christopher Janaway and Simon Robertson, eds., *Nietzsche, Naturalism, and Normativity* (Oxford: Oxford University Press, 2012).

41. Nietzsche, *Götzen-Dämmerung*, KSA, 6:86.

42. This shift is the decisive point of Nietzsche's claim that psychology, rather than logic, metaphysics, or epistemology, is to become the fundamental discipline for resolving philosophical problems, as stated, for example, in *Jenseits von Gut und Böse*, KSA, 5:39.

43. Nietzsche, *Die Geburt der Tragödie*, KSA, 1:16.

44. Nietzsche, *Ecce Homo*, KSA, 6:307.

45. For a philosophical defense of Nietzsche's genealogy as a study of biological *and* cultural selection, see Richardson, *Nietzsche's New Darwinism*.

46. The nature of the evolutionary disposition to morality and its origin is still a major and controversial question for evolutionary theory today. It is currently being addressed through

evolutionary game theory, the empirical study of animal behavior, and continuing debate over different "levels" and "units" of selection that involve the question of the validity of group selection suggested early on by Darwin. For an excellent discussion of the problem of the evolution of morality, see Elliot Sober and David Sloan Wilson, *Unto Others: The Evolution of Altruism* (Cambridge, MA: Harvard University Press, 1998). See also Samir Okasha, *Evolution and the Levels of Selection* (Oxford: Oxford University Press, 2008) for an up-to-date and clear analysis of recent debates over the unit of selection in evolution.

47. Nietzsche, *Genealogie*, KSA, 5:257.

48. Hussain, "Role of Life in the *Genealogy*," 163–65.

49. Nietzsche, *Genealogie*, KSA, 5:252.

50. Ibid., 5:375.

51. See, for example, KSA, 5:265; 5:321; 5:349; 5:352. Looking forward to chapter two, Nietzsche's colleague and friend at the University of Basel, Franz Overbeck, argued much earlier, on the basis of philological investigation of the New Testament, that the ascetic ideal was the "essence" of the early Christian view of life.

52. Nietzsche, *Genealogie*, KSA, 5:366; my emphasis.

53. Ibid., 5:389. In other places in the third essay, Nietzsche describes the affect for which the ascetic ideal provided the cure as "tiredness" (*Müdigkeit*), "heaviness" (*Schwere*), "pain" (*Schmerz*), and "inhibition" (*Hemmung*).

54. Ibid., 5:363.

55. The passage continues as follows: "indeed with this power he binds to existence the entire herd of lost souls, the disgruntled, those who lost out, those who met with an accident, those who suffer from themselves, in that he instinctively leads them as a shepherd" (ibid., 5:366).

56. Ibid., 5:405–6.

57. Ibid., 5:402.

58. Nietzsche's view of the importance of life for ethics as I have articulated it here has important parallels in modern Aristotelian naturalism as put forth, for example, in Philippa Foot's *Natural Goodness* (Oxford: Oxford University Press, 2001).

59. Nietzsche, *Antichrist*, KSA, 6:217.

60. Hussain, "Role of Life in the *Genealogy*," 162.

61. For an illuminating discussion of this problem, see Poellner, *Nietzsche and Metaphysics*, 288.

Chapter Two

1. This name comes from the landlord of the building, Frau Baumann, but it is also a play on words that references a famous ancient cave in central Germany. See Friedrich Nietzsche, Franz Overbeck, and Ida Overbeck, *Friedrich Nietzsche, Franz und Ida Overbeck Briefwechsel*, ed. Katrin Meyer and Barbara von Reibnitz (Stuttgart: J. B. Metzler, 2000). For an account of their relationship written by one of Overbeck's students, see Carl Albrecht Bernoulli, *Franz Overbeck und Friedrich Nietzsche: Eine Freundschaft*, 2 vols. (Jena: Diederichs, 1908).

2. The apt phrase "brothers in arms" comes from Andreas Sommer, *Geist der Historie*. The two main works in English that discuss Overbeck are Martin Henry, *Franz Overbeck: Theologian?* (Frankfurt: Peter Lang, 1995), and Lionel Gossman, *Basel in the Age of Burckhardt: A Study in Unseasonable Ideas* (Chicago: University of Chicago Press, 2000).

3. This phrase comes from Daniel Dennett in *Darwin's Dangerous Idea*, 467.

4. See Andreas Sommer, *Geist der Historie*, 17.

5. For a fascinating study of the fate of F. C. Baur and Hegelian history of religion, see Johannes Zachhuber, *Theology as Science in Nineteenth-Century Germany: From F. C. Baur to Ernst Troeltsch* (Oxford: Oxford University Press, 2013).

6. Gossman, *Basel in the Age of Burckhardt*, 413–38.

7. Overbeck consistently uses the concepts of *Lebensideal* (ideal of life), *Lebensansicht* (view of life), and *Lebensbetrachtung* (perspective on life) to characterize what religions are. See Franz Overbeck, *Ueber die Christlichkeit unserer heutigen Theologie*, in *Franz Overbeck Werke und Nachlass*, vol. 1, ed. Ekkehard W. Stegemann and Niklaus Peter (1994), 238.15–20. Citations to works from the critical edition of Overbeck's works follow the format [page number].[line number].

8. This title is taken from Martin Henry's translation of Overbeck's *Christlichkeit*: Franz Overbeck, *How Christian Is Our Present-Day Theology*, trans. Martin Henry (London: Continuum, 2005). Henry's translation of the title preserves the sense of Overbeck's inquiry as an attempt to judge whether or not academic theology has the right to call itself genuinely "Christian," and thus to judge modern Christian culture in relation to its early historical forms. Throughout the rest of this chapter, all of the translations from the critical edition of Overbeck's works are mine unless otherwise stated.

9. See Martin Henry, "Franz Overbeck: A Review of Recent Literature," *Irish Theological Quarterly* 72 (2007): 391–404. Löwith dedicates the final chapter of his massive history of nineteenth-century thought to Overbeck; see *Von Hegel zu Nietzsche: Der Revolutionäre Bruch im Denken des neunzehnten Jahrhunderts*. (Stuttgart: Kohlhammer, 1950).

10. Theodore Kisiel, *The Genesis of Heidegger's Being and Time* (Berkeley: University of California Press, 1995), 556–57, n. 15.

11. See Niklaus Peter, "Ernst Troeltsh auf der Suche nach Franz Overbeck: Das Problem der Historismus in Perspektive zweier Theologen," *Troeltsch-Studien* 11 (2000): 94–122.

12. See David Tracy's foreword to Martin Henry's English translation of Overbeck's main philosophical work, *How Christian Is Present-Day Theology?* is an appreciative call to contemporary theologians and philosophers of religion to pay more attention to Overbeck's analysis of religion and theology. See also the essays in Rudolf Brändle and Ekkehard Stegemann, eds., *Franz Overbecks unerledigte Anfragen an das Christentum* (Munich: Chr. Kaiser Verlag, 1988.), in particular Niklaus Peter's essay in this volume, "Unerledigte Anfragen und befragte Erledigungen. Eine Erste Rezeption und Diskussion dreier Beiträge," 196–210.

13. Niklaus Peter, *Im Schatten der Modernität: Franz Overbecks Weg zur "Christlichkeit unserer heutigen Theologie"* (Stuttgart: J. B. Metzler, 1992), 9–17.

14. Franz Overbeck, "Über Entstehung und Recht einer rein historischen Betrachtung der Neutestamentlichen Schriften in der Theologie" (1871), in *Franz Overbeck Werke und Nachlass*, vol. 1, ed. Ekkehard W. Stegemann and Niklaus Peter (1994), 97.14–30.

15. For a contemporary interpretation of the teleological character of Hegel's naturalism, see Terry Pinkard, *Hegel's Naturalism: Mind, Nature, and the Final Ends of Life* (Oxford: Oxford University Press, 2012), 24–26.

16. Ibid., 17–33.

17. Overbeck, "Entstehung," 90.20.

18. Ibid., 90.22–25.

19. Ibid., 99.26–35.

20. Ibid., 104.7–14.

21. Bernard Williams, *Truth and Truthfulness: An Essay in Genealogy* (Princeton, NJ: Princeton University Press, 2002), 162–63.

22. I thank Andreas Sommer for this comment on Overbeck's assessment of Eusebius as a historian.

23. Williams, *Truth and Truthfulness*, 161.

24. See Sommer, *Geist der Historie*, 29.

25. Niklaus Peter, *Im Schatten der Modernität*, 239.

26. Overbeck, *Christlichkeit*, 169.2–4. I am using Martin Henry's translation of *Bildung* here from *How Christian?* 7.

27. Overbeck, *Christilichkeit*, 205.32.

28. Paul de Lagarde, *Über das Verhältnis des deutschen Staates zu Theologie, Kirche und Religion: Ein Versuch Nicht-Theologen zu orientieren* (Göttingen: Dieterichsche Verlagsbuchhandlung, 1873) [On the relation of the German state to theology, church, and religion: An attempt to orient non-theologians]. See Niklaus Peter, *Im Schatten der Modernität*, 173–81.

29. Overbeck, *Christlichkeit*, 238.16–20. See Niklaus Peter, *Im Schatten der Modernität*, 182–89.

30. See Sommer, *Geist der Historie*, 6.

31. Overbeck, *Christlichkeit*, 238.15–20.

32. Ibid., 179–80.32–33.

33. Ibid., 180.10–12.

34. Ibid., 205.27–31.

35. Ibid., 180.19.

36. Ibid., 206.21–30.

37. Ibid., 109–113.

38. Ibid., 215.11–15.

39. Ibid., 216.31–36.

40. See Richard Finn, *Asceticism in the Graeco-Roman World* (Cambridge: Cambridge University Press, 2009).

41. For an analysis of Overbeck's reading of Schopenhauer, see Andreas Urs Sommer, "Weltentsagung, Skepsis und Modernitätskritik: Arthur Schopenhauer und Franz Overbeck," *Philosophisches Jahrbuch* 107, no. 1 (2000): 192–206.

42. Overbeck, *Christlichkeit*, 171–72.27–25.

43. Ibid., 172.12–15; emphasis in the original.

44. See Henry, *Franz Overbeck: Theologian?* 189–200.

45. See Gould, "Non-Overlapping Magisteria." Overbeck's account of the reason for the divergence between these two "magisteria" is, of course, very different from Gould's, which distinguished them not by their sources of normativity and divergent interests but by the usual positivist distinction between facts and values.

46. Overbeck, *Christlichkeit*, 183.5–11.

47. Ibid., 181.20–22.

48. See Andreas Sommer, "On the Genealogy of the Genealogical Method: Overbeck, Nietzsche, and the Search for Origins," *Bulletin of the Institute of Classical Studies* Supplement Series 46, supp. 79 (2003): 87–103. Overbeck's historical essays after the *Christlichkeit* ranged from topics such as "On the Relationship of the Ancient Church to Slavery in the Roman Empire" (1875), "On the Summary of the Debate between Paul and Peter in Antioch (Gal. 2, 11ff.) among the Church Fathers" (1877), and "Towards a History of the Canon" (1880), to essays such as "On the Beginnings of Patristic Literature" (1882) and "On the Beginnings of the Writing of Church History" (1892).

49. See Sommer, *Geist der Historie*, 92–94.

50. Franz Overbeck, *Kirchenlexicon Texte: Ausgewählte Artikel A–I*, in *Franz Overbeck Werke und Nachlass*, vol. 4, ed. Barbara von Reibnitz (1995), 270.14–271.32.

51. Overbeck, *Kirchenlexicon A–I*, 601.28–602.1

52. Franz Overbeck, *Kirchenlexicon Texte: Ausgewählte Artikel J–Z*, in *Franz Overbeck Werke und Nachlass*, vol. 5, ed. Barbara von Reibnitz (1995), 299.1–13. Overbeck is quoting here the scene from Goethe's *Faust* in which Mephistopheles describes himself as the "spirit of negation," proclaiming "*denn alles was ensteht / Ist werth, daß es zugrunde geht*" (For everything that comes to be, is fit to perish). Johann Wolfgang von Goethe, *Faust Part I*, trans. David Constantine (London: Penguin Classics, 2005), 46. Nietzsche quotes this verse in his 1874 *Untimely Meditation*, "On the Use and Abuse of History for Life," and in *Thus Spoke Zarathustra*.

53. Overbeck, *Kirchenlexicon: J–Z*, 235.1–18.

54. Ibid., 237.14–17.

55. Franz Overbeck, *Selbstbekenntnisse: Mit einer Einleitung von Jacob Taubes* (Frankfurt am Main: Insel Verlag, 1966), 145–46.

56. Franz Overbeck, *Kirchenlexicon Materialen*, in *Franz Overbeck Werke und Nachlass*, vol. 6, bk. 1, ed. Barbara von Reibnitz (1996), 303.7–22.

57. See Friedrich Wilhelm Graf, "Overbeck, Franz Camille," in *Religion in Geschichte und Gegenwart*, ed. Hans Dieter Betz et al., vol. 6 (Tübingen: Mohr Siebeck, 2003), 758–59. Ernst Troeltsch also makes this point in his review of the *Christlichkeit*. See his "Review of Franz Overbeck: *Über die Christlichkeit unserer Heutigen Theologie*," in *Rezensionen und Kritiken (1901–1914)*, vol. 4 of *Ernst Troeltsch Kritische Gesamtausgabe*, ed. Friedrich Wilhelm Graf (Berlin: Walter de Gruyter, 2004), 292–95.

Chapter Three

1. Georg Simmel, *Soziologie*, in *Georg Simmel Gesamtausgabe*, vol. 11, ed. Otthein Rammstedt (Frankfurt am Main: Suhrkamp, 1992), 13–62.

2. Beiser, *German Historicist Tradition*, 472–74.

3. Beiser shows Simmel's early work to be a product of the same intellectual milieu of the aftermath of Hegel's idealism that is also a characteristic of Nietzsche, Overbeck, and, of course, the forefather of *Lebensphilosophie*, Schopenhauer. See Beiser, *German Historicist Tradition*, 472–74.

4. Ibid., 485–510.

5. This is only one of various tripartite schemata that have been offered to account for Simmel's development. These schemata are useful, but, as recent scholars have noted, they give a false impression of discontinuity in Simmel's concerns throughout these periods. See the helpful discussions of the periodization of Simmel's work in Hans Joas, *The Genesis of Values*, trans. Gregory Moore (Chicago: University of Chicago Press, 2000), 70. See also Volkhard Krech, *Georg Simmels Religionstheorie*, Religion und Aufklärung, vol. 4 (Tübingen: Mohr Siebeck, 1998), 8–10.

6. Klaus Christian Köhnke, *Der junge Simmel* (Frankfurt: Suhrkamp, 1996), 10.

7. Simmel, *Georg Simmel Gesamtausgabe*, vol. 10, ed. Micheal Behr, Volkhard Krech, and Gert Schmidt (1995), 419–20.

8. See Frederick Beiser's analysis of Simmel's *Einleitung in die Moralwissenschaft* in *German Historicist Tradition*, 502–10.

9. Simmel's early essays on Nietzsche include "Friedrich Nietzsche: Eine Moralphilosophische Silhouette" (1896); "Zum Verständnis Nietzsches" (1902); and "Nietzsche und Kant" (1906). He

also published a short excerpt from his monograph that he also titled *Schopenhauer und Nietzsche* separately and a year earlier in 1906, which he claimed drew from his 1902 lectures in Berlin.

10. For an excellent set of contemporary essays on Nietzsche's relationship to Schopenhauer, see Christopher Janaway, ed. *Willing and Nothingness: Schopenhauer as Nietzsche's Educator* (Oxford: Clarendon Press, 1998).

11. An example of the influence of Simmel's interpretation of Nietzsche on intellectuals in his time is the fact that Ernst Troeltsch credited his interpretation of Nietzsche to Simmel's writings and based his treatment of Nietzsche on them in *Historicism and Its Problems*. See Friedemann Voigt, *Die Tragödie des Reiches Gottes?: Ernst Troeltsch als Leser Georg Simmels*, Troeltsch-Studien, vol. 10 (Gütersloh: Gütersloher Verlag Haus, 1998), 58.

12. Georg Simmel, *Schopenhauer und Nietzsche*, in *Georg Simmel Gesamtausgabe*, vol. 10, ed. Micheal Behr, Volkhard Krech, and Gert Schmidt (1995), 386.

13. This view is still accepted in Nietzsche scholarship today. See Janaway, *Willing and Nothingness*, 1–12.

14. Simmel, *Schopenhauer und Nietzsche*, 178. All translations in this chapter are my own unless otherwise specified.

15. Ibid.

16. Ibid., 179; emphasis in the original.

17. See R. Lanier Anderson, "Neo-Kantianism and the Roots of Anti-Psychologism," *British Journal of the History of Philosophy* 13, no. 2 (2005): 287–323.

18. Simmel, *Schopenhauer und Nietzsche*, 386.

19. Ibid., 387.

20. Georg Simmel, *Die Philosophie des Geldes*, in *Georg Simmel Gesamtausgabe*, vol. 6, ed. David Frisby and Klaus Christian Köhnke (1989), 12.

21. Ibid., 12.

22. Ibid., 246–53.

23. Ibid., 13.

24. Ibid., 10.

25. Ibid., 24; my emphasis.

26. I take this phrase from the title of Thomas Nagel's book, which reflects on similar issues. See *The View from Nowhere* (Oxford: Oxford University Press, 1989).

27. The reference to Spinoza on this point of the unity of the theoretical and the practical recurs throughout Simmel's corpus. Simmel's monistic point here resonates with contemporary philosophical defenses of "neutral monism" in response to the mind/body problem. This position argues that the world is ultimately made up of one sort of "stuff," but this is neither physical nor mental. See John Heil, *Philosophy of Mind: A Contemporary Introduction*, 3rd ed. (New York: Routledge, 2013), 241.

28. Simmel, *Philosophie des Geldes*, 29.

29. Ibid., 41.

30. Ibid.

31. Ibid., 43.

32. Ibid., 54.

33. Ibid., 59.

34. Georg Simmel, *Lebensanschauung* (Munich: Duncker & Humboldt, 1918), 20.

35. Ibid.

36. Ibid.

37. An important intermediary figure between Nietzsche and Simmel who contributed to

NOTES TO CHAPTER FOUR　　　　　　　　　　　　　　　　　　　　　165

Simmel's retooled concept of life was the French vitalist Henri Bergson, about whom Simmel also wrote and commented. See Georg Simmel, "Henri Bergson (1914)," in *Georg Simmel Gesamtausgabe*, vol. 13, ed. Klaus Latzel (1992), 53–69. Simmel's Life philosophy as a whole could be called an attempt to synthesize what he saw as the most significant aspects of Nietzsche, Darwin, Bergson, and Kant.

38. Georg Simmel, *Lebensanschauung*, 24.

39. Content (*Gehalt*) at the level of the intellect involves, for Simmel, symbols, logic, values, representations, concepts, and anything that belongs to the domain of meaning and the mind.

40. Georg Simmel, *Lebensanschauung*, 161–62.

41. This passage also, of course, recalls Nietzsche's description of life as "the will to life rejoicing over its own inexhaustibility, even in the very sacrifice of its highest types" in *Ecce Homo*, KSA, 6:312

42. Georg Simmel, "Beiträge zur Erkenntnistheorie der Religion," in *Georg Simmel Gesamtausgabe*, vol. 7, ed. Rüdiger Krämme, Angela Rammstedt, and Otthein Rammstedt, 9–10.

43. Ibid., 18.

44. Ibid.

45. Georg Simmel, "Die Religion," in *Georg Simmel Gesamtausgabe*, vol. 10, ed. Michael Behr, Volkhard Krech, and Gert Schmidt (1995), 48.

46. Ibid., 46.

47. Simmel, *Lebensanschauung*, 57–58.

48. Ibid., 58.

49. For a helpful discussion of this concept of a "series of purposes," see Krech, *Georg Simmel Religionstheorie*, 87–91.

50. Simmel, *Lebensanschauung*, 87–88.

51. Ibid., 58.

52. See Simmel, *Philosophie des Geldes*, 297. Simmel is here weighing in on a significant debate in political economy that included Marx, Weber (with whom Simmel shared a personal friendship and professional exchange), and Ferdinand Tönnies over the breakdown of traditional forms of community in modern industrial societies.

53. These themes are worked out in the second, "synthetic," part of *Die Philosophie des Geldes*, 283–302.

54. Simmel, *Lebensanschauung*, 92–96.

55. Georg Simmel, "Der Begriff und die Tragödie der Kultur," in *Georg Simmel Gesamtausgabe*, vol. 12, ed. Rüdiger Kramme and Angela Rammstedt (2001), 496. Simmel originally published this essay in 1911.

56. See Simmel, *Lebensanschauung*, 154.

57. Ibid., 26.

Chapter Four

1. Georg Simmel, *Lebensanschauung*, 26.

2. See Peter Woodford, "The Very Possibility of a Science of Religion: Ernst Troeltsch and Neo-Kantianism," *Journal of Religion* 97, no. 1 (2017): 56–78.

3. See Cassirer's two volumes on the philosophy of symbolic forms: *Philosophie der symbolischen Formen. Erster Teil: Die Sprache* (Berlin: Bruno Cassirer 1923); and *Philosophie der symbolischen Formen. Zweiter Teil: Das mythische Denken*, (Berlin: Bruno Cassirer 1925).

4. Rickert does not mention Franz Overbeck in his book, although his criticisms of Life-

philosophy in general show that he equally belongs in Rickert's discussion. The reason for Rickert's neglect is that Overbeck had less impact on philosophical circles than Nietzsche or Simmel.

5. Rickert's early historical account comes to many of the same conclusions as recent historical studies of Life-philosophy. See, for example, Jürgen Größe, *Lebensphilosophie* (Stuttgart: Reclam Verlag, 2010).

6. Heinrich Rickert, *Die Philosophie des Lebens: Darstellung und Kritik der Philosophischen Modeströmungen unserer Zeit* (Tübingen: J. C. B. Mohr, 1920), 145. All translations are my own unless otherwise specified.

7. Ibid., iv.

8. Ibid., iv–v.

9. Ibid., 7.

10. I will continue to use the term *biologism* to describe the school of thought that Rickert is criticizing, but its relation to contemporary concerns over naturalism should be kept in mind. *Biologism*, like *naturalism*, refers to the attempts to provide a comprehensive account of the basis of human values in non- or pre-rational affects and processes operative throughout the non-human, living world more generally. See Rickert, *Philosophie des Lebens*, 7–11.

11. See Tim Lewens, *Darwin* (New York: Routledge, 2007), 26.

12. Rickert, *Philosophie des Lebens*, 79. Given the contemporary surge of popular and academic interest in evolutionary theories of religion, Rickert's worries here are of more than historical interest.

13. Rickert also mentions William James in connection with Simmel's pragmatist conception of truth and knowledge. Rickert's lumping together of Life-philosophy and pragmatism presents an interesting example of early views of the relation between American pragmatism and post-Kantian German thought.

14. Rickert's critique of intuitionism is more directed against Bergson and Dilthey than the "Nietzschean" strand of Life-philosophy, which did not inherit this romantic privileging of intuition over conceptual thought. Overbeck has often been interpreted this way, but as I argued in chapter 2, his distinction between science and life ought not to be confused with that between "concept" and "feeling." It is best understood as a conflict between the aims of scientific knowledge and those of a religious "way of life" oriented toward the value of life itself.

15. Rickert uses the term "monism" throughout his work to characterize the central ambition of Life-philosophy (Rickert, *Philosophie des Lebens*, 89).

16. Of course, in arguing this, Rickert was rejecting pan-organicist and pan-vitalist metaphysical views that can be traced back to Leibniz and were present in his day, notably in the influential writings of Rudolf Eucken. I thank Brent Sockness for pointing out the importance of Eucken to these discussions.

17. Rickert, *Philosophie des Lebens*, 105.

18. Ibid..

19. Ibid.

20. Rickert's thought offers an interesting example of how American pragmatism was received by thinkers in the Idealist tradition of German philosophy. He viewed pragmatism as a species of naturalism, along with Life-philosophy, that succumbed to the same problems as the Nietzscheans and so-called "traditional" Darwinists.

21. See Moore, *Nietzsche, Biology, and Metaphor*. Rickert's point here is controversial and raises important problems in contemporary philosophy of biology regarding the normativity of functional ascriptions and the kinds of normative claims that can be made on the basis of the

proper functioning of organisms and their parts. I discuss some of these issues in relation to contemporary debates about natural normativity in the conclusion.

22. Rickert, *Philosophie des Lebens*, 126; my emphasis.

23. Ibid., 128. Of course, as we saw, it remains questionable what sort of normativity Nietzsche wanted to grant his own stance of life-affirmation. It is clear that he would be happy to reject the kind of *rational* necessity that Rickert insists on and instead to rest his values on the biological necessity from which they originate. Rickert would worry that this would sacrifice the self-sufficiency and self-reliance of rationality that could redeem the validity of this value. I return to this crucial issue in the conclusion.

24. Ibid.

25. Ibid., 129.

26. Ibid., 128.

27. Ibid., 132–33.

28. Ibid., 131.

29. Ibid., 141.

30. Ibid., 151.

31. Ibid., 157.

32. Ibid., 189.

33. Ibid., 169.

34. Heinrich Rickert, *Der Gegenstand der Erkenntnis: Einführung in die Transzendentalphilosophie*, 5th ed. (Tübingen: Mohr-Siebeck), 1921. The list of interlocutors against whom Rickert defended his Kantian epistemology included founder of phenomenology Edmund Husserl, theorist of the human sciences and philosopher of history Wilhelm Dilthey, and founder of empirical psychology Wilhelm Wundt.

35. Rickert's critique of psychologism also targets Dilthey and a collection of psychologists and philosophers of psychology, including the founder of empirical psychology, Wilhelm Wundt. See Anderson, "Neo-Kantianism."

36. Another difference between the Marburg and Southwest schools lay in their theories of the relation between intuition and concept. Rickert held that grasping experience in concepts always requires an act of selection that leaves some aspects of experience out. Thus, Rickert emphasized the fact that experience always exceeds conceptual grasp in a way that the Marburgers did not.

37. Rickert, *Gegenstand*, 380.

38. Ibid., 190.

39. Ibid., 121.

40. Ibid., 271.

41. This sort of strategy is alive and well in philosophy today. See Thomas Nagel's similar use of this form of transcendental argument to refute skepticism, historicism, and relativism in *The Last Word* (Oxford: Oxford University Press, 1997).

42. Rickert, *System der Philosophie*, 120.

43. Ibid., 113.

44. Ibid., 114; my emphasis.

45. Heinrich Rickert, "Vom Begriff der Philosophie (1910)," in *Philosophische Aufsätze*, ed. Rainer A. Bast (Tübingen: Mohr Siebeck, 1999), 8.

46. Ibid., 8.

47. Rickert often uses both "meaning" [*Sinn*] and "value" interchangeably and views intelligible "meaning" also as involving the instantiation of "values." See Heinrich Rickert, "Die

Methode der Philosophie und das Unmittelbare (1924)," in *Philosophische Aufsätze*, ed. Rainer A. Bast (Tübingen: Mohr Siebeck, 1999), 140–51.

48. Rickert, "Vom Begriff," 11; emphasis in the original.

49. Ibid., 12.

50. For a competing *psychological* approach to the problem of "worldview" that opposed Rickert and that Rickert also criticized, see Karl Jaspers, "Psychologie der Weltanschauungen und Philosophie der Werte," in *Logos* 9 (1920/21): 1–42.

51. Rickert, *System der Philosophie*, 150.

52. One of these texts is a monograph published in 1924 entitled *Kant als Philosoph der Modernen Kultur: Ein Geschichtsphilosophischer Versuch* (Tübingen: J. C. B. Mohr, 1924). This is a fascinating text that applies some of the points Rickert makes in his general system of values to argue for a Kantian conception of modernity in terms of the differentiation and disunity of the spheres of culture.

53. See Rickert, *System der Philosophie*, 356.

54. Rickert further classifies these three "future goods" [*Zukunftsgüter*], "present goods" [*Gegenwartsgüter*], and "eternal goods" [*Ewigkeitsgüter*] respectively. See Rickert, *System der Philosophie*, 379–80.

55. Ibid., 340–41.

56. Rickert's notion of religion as a distinct value sphere that strives to realize the transcendental value of "holiness" is similar to Simmel's theory of religion presented in the last chapter. Of course, Rickert sees the origin of this value in the rational structure of agency and not in life.

57. Rickert, *System der Philosophie*, 340.

58. Ibid., 340–41.

59. Rickert defines a "church," or the religious community, as the "care" (*Pflege*) for the religious ideal in social life (ibid., 341).

60. Ibid., 340.

61. For a helpful analysis and reconstruction of Rickert's theory of religion, see Benjamin Crowe, "Faith and Value: Heinrich Rickert's Theory of Religion," in *Journal of the History of Ideas* 71, no. 4 (2010): 617–36.

62. Rickert, *System der Philosophie*, 340.

63. Rickert's student Emil Lask was one of his most important critics on this point. He challenged Rickert's dualism by offering an ontological theory of judgment that later influenced Martin Heidegger. See Emil Lask, *Die Lehre Vom Urteil* (Tübingen: J. C. B. Mohr, 1912). Ernst Troeltsch also criticized Rickert on this point.

64. Frederick Beiser shares this critical estimation of Rickert's project as a failure because of its inability to show how *Sollen* and *Sein* could be unified, but he does not consider Rickert's philosophy of religion and notion of a basic metaphysical "faith" with respect to this issue. See Beiser, *German Historicist Tradition*, 439–41.

65. These late essays include "Thesen zum System der Philosophie" (1932), "Wissenschaftliche Philosophie und Weltanschauung" (1933), and "Die Heidelberger Tradition und Kants Kritizismus" (1934).

66. Heinrich Rickert, "Die Heidelberger Tradition und Kants Kritizismus (Systematische Selbstdarstellung," in *Philosophische Aufsätze*, ed. Rainer A. Bast (Tübingen: Mohr-Siebeck, 1999), 401.

67. Ibid., 401.

68. Ibid.

69. Ibid., 403–5.

70. Heinrich Rickert, "Fichtes Atheismusstreit und die Kantische Philosophie: Eine Säkularbetrachtung," *Kantstudien* 4, nos. 1–3 (1899): 137–66.

Conclusion

1. Nietzsche's concept of life's basic affirmation of itself as the ground of value and science has interesting parallels with the concept of a basic "animal faith" in the work of George Santayana. See George Santayana, *Skepticism and Animal Faith* (New York: Dover Press, 1955).

2. Two classic studies of this debate that offer interpretations of its significance for Continental and Analytic philosophy are by Michael Friedman, *A Parting of the Ways: Carnap, Cassirer, Heidegger* (Chicago: Open Court Publishing, 2000); and Peter E. Gordon, *Continental Divide: Heidegger, Cassirer, Davos* (Cambridge, MA: Harvard University Press, 2010).

3. See Jürgen Habermas, *Erkenntnis und Interesse* (Frankfurt am Main: Suhrkamp, 1994); see also the transcendental argument for discourse ethics in Karl-Otto Apel, *Understanding and Explanation: A Transcendental-Pragmatic Perspective*, trans. Georgia Warnke (Cambridge, MA: MIT Press, 1984).

4. For a fascinating study of Heidegger's relationship to Life-philosophy, see David Farrell Krell, *Daimon Life: Heidegger and Life-Philosophy* (Bloomington: Indiana University Press, 1992).

5. Marjorie Grene, *The Understanding of Nature: Essays in the Philosophy of Biology* (Boston: Dordrecht, 1974), 172.

6. For a helpful overview of contemporary discussions of teleology, see Ariew, Cummins, and Perlman, *Functions*. For a powerful critique of the mechanistic, gene-centered focus of biology, see Walsh, *Organisms, Agency, and Evolution*.

7. There are many diverse strands of contemporary debate over evolution and ethics that span the natural sciences, social sciences, philosophy, and the study of religion. For a glimpse of two sides that distill many of the major meta-ethical issues, see the disagreement between Sharon Street and Thomas Nagel cited in the introduction. See Street, "Darwinian Dilemma"; and Nagel, *Mind and Cosmos*, 97. Hannah Ginsborg has also offered a compelling recent interpretation and defense of Kant's philosophy of biology in *The Normativity of Nature: Essays on Kant's Critique of Judgment* (Oxford: Oxford University Press, 2014).

8. See, for example, the discussion of the concept of life in Kim Sterelny and Paul Griffiths, *Sex and Death: An Introduction to Philosophy of Biology* (Chicago: University of Chicago Press, 1999); Peter Godfrey-Smith, *Philosophy of Biology* (Princeton: Princeton University Press, 2014); and a recent textbook on evolution that defines life in terms of metabolism and replication, Stephen C. Stearns, and Rolf F. Hoekstra, *Evolution: An Introduction*, 2nd ed. (Oxford: Oxford University Press, 2005), 356.

9. The legitimacy of the agential picture of organisms is itself the subject of fascinating debate within the philosophy of biology. See Peter Godfrey-Smith, *Darwinian Populations and Natural Selection* (Oxford: Oxford University Press, 2009), 9–15.

10. See Nagel, *Mind and Cosmos*, 92.

11. For debates over neo-Aristotelian positions on these issues, see Michael Thompson, *Life and Action: Elementary Structures of Practice and Practical Thought* (Cambridge, MA: Harvard University Press, 2008); and Philippa Foot, *Natural Goodness* (Oxford: Clarendon Press, 2001). For a discussion of how these Aristotelian conceptions of life relate to neo-Darwinian discussions of teleology, see Peter Woodford, "Neo-Darwinism and Neo-Aristotelianism: How to Talk about Natural Purpose," *History and Philosophy of the Life-Sciences* 38, no. 4 (2016): 1–23.

12. Foot dedicates the last chapter of *Natural Goodness* to Nietzsche. For a critique of Thompson and Foot along Nietzschean lines that accepts the account of natural values in biological teleology, see Chrisoula Andreou, "Getting On in a Varied World," *Social Theory and Practice* 32, no. 1 (2006): 61–73; and Elijah Millgram, critical notice of *Life and Action*, *Analysis Reviews* 69, no. 3 (2009): 557–64.

13. For criticisms of Thompson and Foot that may be more aligned with a neo-Kantian view of biology and ethics, see William Fitzpatrick, *Teleology and Norms of Nature* (New York: Routledge, 2011); and Tim Lewens, "Human Nature: The Very Idea," *Philosophy and Technology* 25 (2012): 459–74.

14. It must be noted that Weber too argued, thanks to his reception of Rickert and Nietzsche, that science aimed at the value of truth *for its own sake* and was not value-free. See Max Weber, "Science as a Vocation," in *The Vocation Lectures*, ed. David Owen and Tracy Strong, trans. Rodney Livingstone (Indianapolis, IN: Hackett, 2004).

15. My aim here is not to argue for this as the sole or even best definition of religion, but only to affirm that it is as an acute observation of an aspect of many cultural forms that are commonly called religions that is of particular philosophical interest. The question of the definition of religion in the academic study of religion is as difficult as the definition of life in biology.

16. Many philosophers today turn to pragmatism as a way of answering transcendental objections while avoiding metaphysics and remaining committed to naturalism. See, for example, Hilary Putnam, *Ethics without Ontology* (Cambridge, MA: Harvard University Press, 2004).

17. For a contemporary defense of this sort of view, see Robert Pippin, "Natural and Normative," *Daedelus* 138, no. 3 (2009): 35–43.

18. For an example of a recent work that can be challenged by these views, see Edward Slingerland, *What Science Offers the Humanities: Integrating Body and Culture* (Cambridge: Cambridge University Press, 2008). Slingerland accepts physicalism as the default metaphysical basis of the sciences and does not address questions about sources of normativity and validity. He does not address meta-ethical and metaphysical debates that challenge physicalism and question the limits of scientific explanations of religion.

19. For philosophical approaches consistent with the meta-ethical and metaphysical focus here, see Ronald Dworkin, *Religion without God* (Cambridge, MA: Harvard University Press, 2013); Thomas Nagel, *Secular Philosophy and the Religious Temperament: Essays 2002–2008* (Oxford: Oxford University Press, 2010); Mark Johnston, *Saving God: Religion After Idolatry* (Princeton, NJ: Princeton University Press, 2009); and Franklin Gamwell, *Existence and the Good: Metaphysical Necessity in Morals and Politics* (Albany: SUNY Press, 2011).

20. For a recent, positivist, account of the relationship between religion and science along these lines that has been influential in the sciences, see Coyne, *Faith vs. Fact*.

Bibliography

Primary Literature

Barth, Karl. "Unerledigte Anfragen an die Heutige Theology (1920)." In *Karl Barth: Die Theologie und die Kirche. Gesammelte Vorträge*, 2:1–25. Munich: C. Kaiser, 1928.

Bernoulli, Carl Albrecht. *Franz Overbeck und Friedrich Nietzsche: Eine Freundschaft*. 2 vols. Jena: Diederichs, 1908.

Burckhardt, Jacob. *History of Greek Culture*. Translated by Palmer Hilty. New York: Dover Press, 2002.

Cassirer, Ernst. *Philosophie der symbolischen Formen. Erster Teil: Die Sprache*, Berlin: Bruno Cassirer, 1923.

———. *Philosophie der symbolischen Formen. Zweiter Teil: Das mythische Denken*, Berlin: Bruno Cassirer, 1925.

Goethe, Johann Wolfgang von. *Faust, Part I*. Translated by David Constantine. London: Penguin Classics, 2005.

Jaspers, Karl. "Psychologie der Weltanschauungen und Philosophie der Werte." *Logos* 9 (1920–21): 1–42.

Kant, Immanuel. *The Critique of Pure Reason*. Edited and translated by Paul Guyer and Allen Wood. Cambridge: Cambridge University Press, 1997.

———. *Critique of the Power of Judgment*. Edited by Paul Guyer. Translated by Paul Guyer and Eric Matthews. Cambridge: Cambridge University Press, 2002.

Lagarde, Paul de. *Über das Verhältnis des deutschen Staates zu Theologie, Kirche und Religion: Ein Versuch Nicht-Theologen zu orientieren*. Göttingen: Dieterichsche Verlagsbuchhandlung, 1873.

Lask, Emil. *Die Lehre Vom Urteil*. Tübingen: J. C. B. Mohr, 1912.

Nietzsche, Friedrich. *Sämtliche Werke: Kritische Studienausgabe in 15 Bänden*. Edited by Giorgio Colli and Mazzino Montinari. 2nd ed. 15 vols. Berlin: De Gruyter, 1988.

———, Franz Overbeck, and Ida Overbeck. *Friedrich Nietzsche, Franz und Ida Overbeck Briefwechsel*. Edited by Katrin Meyer and Barbara von Reibnitz. Stuttgart: J. B. Metzler, 2000.

Overbeck, Franz. *Franz Overbeck Werke und Nachlass*. Edited by Niklaus Peter, Ekkehard Stegemann, and Barbara von Reibnitz. 9 vols. Stuttgart: J. B. Metzler, 1994–2010. (Citations in the notes of particular volumes give the editors and publication dates of those volumes.)

———. *How Christian Is Our Present-Day Theology?* Translated by Martin Henry. London: Continuum, 2005.

———. *Selbstbekenntnisse.* Frankfurt am Main: Insel Verlag, 1966.

Rickert, Heinrich. *Der Gegenstand der Erkenntnis: Einführung in die Tranzendental Philosophie.* 5th ed. Tübingen: J. C. B. Mohr, 1921.

———. *Die Philosophie des Lebens: Darstellung und Kritik der Philosophischen Modeströmungen unserer Zeit.* Tübingen: J. C. B. Mohr, 1920.

———. "Fichtes Atheismusstreit und die Kantische Philosophie: Eine Säkularbetrachtung." *Kantstudien* 4, nos. 1–3 (1899): 137–66.

———. *The Limits of Concept Formation in Natural Science: A Logical Introduction to the Historical Sciences.* Edited and translated by Guy Oakes. Cambridge: Cambridge University Press, 1986.

———. *Philosophische Aufsätze.* Edited by Rainer A. Bast. Tübingen: Mohr-Siebeck, 1999.

———. "Psychologie der Weltanschauungen und Philosophie der Werte." *Logos* 9 (1920–21): 1–42.

———. *System der Philosophie: Allgemeine Grundlegung der Philosophie.* Tübingen: Mohr-Siebeck, 1921.

Scheler, Max. "Versuch einer Philosophie des Lebens." In *Gesammelte Werke*, 3:311–41. 15 vols. Bern: Francke, 1955.

Simmel, Georg. *Georg Simmel Gesamtausgabe.* Edited by Otthein Rammstedt. 24 vols. Frankfurt am Main: Suhrkamp, 1992–2016.

———. *Lebensanschauung.* Munich: Duncker & Humboldt, 1918.

———. "On the Concept and Tragedy of Culture." In *Georg Simmel: The Conflict in Modern Culture and Other Essays*, translated by K. Peter Etzkorn. New York: Teachers College Press, 1968.

———. "On the Psychology of Money." In *Simmel on Culture*, edited and translated by Mark Ritter and David Frisby, 233–42. London: Sage, 1997.

———. *The View of Life.* Translated by John A. Y. Andrews and Donald N. Levine. Chicago: University of Chicago Press, 2010.

Strauss, David Friedrich. *Der Alte und Der Neue Glaube: Ein Bekenntniss.* 2nd ed. Leipzig: Hirzel, 1872.

Troeltsch, Ernst. "Die Krisis des Historismus." In *Ernst Troeltsch Kritische Gesamtausgabe*, edited by Gangolf Hübinger, 15:433–56. 20 vols. Berlin: Walter De Gruyter, 2002.

———. "Empiricism and Platonism in the Philosophy of Religion." *Harvard Theological Review* 5, no. 4 (1912): 401–23.

———. "Review of Franz Overbeck, *Über die Christlichkeit unserer Heutigen Theologie*." In *Rezensionen und Kritiken (1901–1914)*, vol. 4 of *Ernst Troeltsch Kritische Gesamtausgabe*, edited by Friedrich Wilhelm Graf, 292–95. Berlin: Walter de Gruyter, 2004.

Secondary Literature

Abbot, Patrick, Jun Abe, John Alcock, Samuel Alizon, Joao A. C. Alpendrinha, Malte Andersson, Jean-Baptiste Andre, et al. "Inclusive Fitness Theory and Eusociality." *Nature* 471 (March 2011): E9–E10.

Anderson, R. Lanier. "Neo-Kantianism and the Roots of Anti-Psychologism." *British Journal of the History of Philosophy* 13, no. 2 (2005): 287–323.

BIBLIOGRAPHY

Andreou, Chrisoula. "Getting On in a Varied World." *Social Theory and Practice* 32, no. 1 (2006): 61–73.

Ariew, Roger, Robert Cummins, and Mark Perlman, eds. *Functions: New Essays in the Philosophy of Psychology and Biology*. Oxford: Oxford University Press, 2002.

Asad, Talal. *Formations of the Secular: Christianity, Islam, Modernity*. Stanford, CA: Stanford University Press, 2003.

———. *Genealogies of Religion: Discipline and Reasons of Power in Christianity and Islam*. Baltimore: Johns Hopkins University Press, 1993.

Bailey, Tom. "Nietzsche the Kantian?" In *The Oxford Handbook of Nietzsche*, edited by Ken Gemes and John Richardson, 134–59. Oxford: Oxford University Press, 2013.

Beiser, Frederick. *The German Historicist Tradition*. Oxford: Oxford University Press: 2011.

———. *Late German Idealism: Trendelenburg and Lotze*. Oxford: Oxford University Press, 2013.

Brändle, Rudolf, and Ekkehard Stegemann, eds. *Franz Overbecks unerledigte Anfragen an das Christentum*. (Munich: Chr. Kaiser Verlag, 1988).

Brandom, Robert. *Tales of the Mighty Dead: Historical Essays on the Metaphysics of Intentionality*. Cambridge, MA: Harvard University Press, 2002.

Brook, John Hedley. "Darwin and Victorian Christianity." In *The Cambridge Companion to Darwin*, 197–218. Cambridge: Cambridge University Press, 2009.

Campioni, Guiliano, Paolo D'Iorio, Maria Cristina Fornari, Francesco Fronterotta, and Andrea Orsucci, eds. *Nietzsches Persönliche Bibliothek*. Berlin: Walter de Gruyter, 2003.

Clark, Maudemarie. *Nietzsche on Truth and Philosophy*. Cambridge: Cambridge University Press, 1990.

———, and David Dudrick. "Nietzsche and Moral Objectivity: The Development of Nietzsche's Metaethics." In *Nietzsche and Morality*, edited by Brian Leiter and Niel Sinhababu, 192–226. Oxford: Clarendon Press, 2007.

Coyne, Jerry. *Faith vs. Fact: Why Science and Religion Are Incompatible*. New York: Penguin Press, 2015.

Crespi, Bernard, and Kyle Summers. "Inclusive Fitness Theory for the Evolution of Religion." *Animal Behavior* 92 (June 2014): 313–23.

Crowe, Benjamin. "Faith and Value: Heinrich Rickert's Theory of Religion." *Journal of the History of Ideas* 71, no. 4 (2010): 617–36.

Dennett, Daniel. *Darwin's Dangerous Idea: Evolution and the Meanings of Life*. New York: Simon & Schuster, 1996.

Dole, Andrew. *Schleiermacher on Religion and the Natural Order*. Oxford: Oxford University Press, 2010.

Dworkin, Ronald. *Religion without God*. Cambridge, MA: Harvard University Press, 2013.

Emden, Christian. *Nietzsche's Naturalism: Philosophy and the Life-Sciences in the Nineteenth Century*. Cambridge: Cambridge University Press, 2014.

Finn, Richard. *Asceticism in the Graeco-Roman World*. Cambridge: Cambridge University Press, 2009.

Fitzpatrick, William. *Teleology and Norms of Nature*. New York: Routledge, 2011.

Foot, Philippa. *Natural Goodness*. Oxford: Clarendon Press, 2001.

Friedman, Michael. *A Parting of the Ways: Carnap, Cassirer, Heidegger*. Chicago: Open Court, 2000.

Gamwell, Franklin. *Existence and the Good: Metaphysical Necessity in Morals and Politics*. Albany: SUNY Press, 2011.

Gemes, Ken, and John Richardson, eds. *The Oxford Handbook of Nietzsche*. Oxford: Oxford University Press, 2013.

Gerhard, Volker. *Die Funken des Freien Geistes*. Edited by Jan Christopher Heilinger and Nikolaos Louidelis. Berlin: Walter de Gruyter, 2011.

Ginsborg, Hannah. *The Normativity of Nature: Essays on Kant's Critique of Judgment*. Oxford: Oxford University Press, 2014.

Glick, Thomas F., ed. *The Comparative Reception of Darwinism*. Chicago: University of Chicago Press, 1988.

Godfrey-Smith, Peter. *Darwinian Populations and Natural Selection*. Oxford: Oxford University Press, 2009.

———. *Philosophy of Biology*. Princeton, NJ: Princeton University Press, 2014.

Gordon, Peter E. *Continental Divide: Heidegger, Cassirer, Davos*. Cambridge, MA: Harvard University Press, 2010.

Gossman, Lionel. *Basel in the Age of Burckhardt: A Study in Unseasonable Ideas*. Chicago: University of Chicago Press, 2000.

Gould, Stephen Jay. "Non-Overlapping Magisteria." *Natural History* 106 (March 1997): 16–22.

Graf, Friedrich Wilhelm. "Overbeck, Franz Camille." In *Religion in Geschichte und Gegenwart*, edited by Hans Dieter Betz, Don S. Browning, Bernd Janowski, and Eberhard Jüngel, 6:758–59. 8 vols. Tübingen: Mohr Siebeck, 2003.

Grene, Marjorie. *The Understanding of Nature: Essays in the Philosophy of Biology*. Boston: Dordrecht, 1974.

Große, Jürgen. *Lebensphilosophie*. Stuttgart: Reclam Verlag, 2010.

Habermas, Jürgen. *Erkenntnis und Interesse*. Frankfurt am Main: Suhrkamp, 1994.

Hadot, Pierre. *Philosophy as a Way of Life: Spiritual Exercises from Socrates to Foucault*. Translated by Michael Chase. Oxford: Wiley-Blackwell, 1995.

Heil, John. *Philosophy of Mind: A Contemporary Introduction*. 3rd ed. New York: Routledge, 2013.

Henry, Martin. *Franz Overbeck: Theologian?* Frankfurt: Peter Lang, 1995.

———. "Franz Overbeck: A Review of Recent Literature." *Irish Theological Quarterly* 72 (2007): 391–404.

Hill, Kevin R. *Nietzsche's Critiques: The Kantian Foundations of His Thought*. Oxford: Oxford University Press, 2003.

Hoffman, David Marc. *Zur Geschichte des Nietzsche-Archivs: Chronik, Studien und Dokumente*. Vol. 2 of *Supplementana Nietzscheana*, edited by Wolfgang Müller-Lauter and Karl Pestalozzi. Berlin: Walter de Gruyter: 1991.

Hussain, Nadeem. "Honest Illusion: Valuing for Nietzsche's Free Spirits." In *Nietzsche and Morality*, edited by Brian Leiter and Neil Sinhababu, 157–91. Oxford: Clarendon Press, 2007.

———. "Nietzsche's Positivism." *European Journal of Philosophy* 12, no. 3 (2004): 326–68.

———. "The Role of Life in the *Genealogy*." In *The Cambridge Critical Guide to Nietzsche's "On the Genealogy of Morality,"* edited by Simon May, 142–69. Cambridge: Cambridge University Press, 2011.

Jablonka, Eva, and Marion Lamb. *Evolution in Four Dimensions: Genetic, Epigenetic, Behavioral, and Symbolic Variation in the History of Life*. Cambridge, MA: MIT Press, 2005.

Janaway, Christopher. "Naturalism and Genealogy." In *A Companion to Nietzsche*, edited by Keith Ansell Pearson, 337–52. Oxford: Blackwell, 2006.

———, ed. *Willing and Nothingness: Schopenhauer as Nietzsche's Educator*. Oxford: Clarendon Press, 1998.

———, and Simon Robertson, eds. *Nietzsche, Naturalism, and Normativity*. Oxford: Oxford University Press, 2012.
Joas, Hans. *The Creativity of Action*. Chicago: University of Chicago Press, 1996.
———. *The Genesis of Values*. Chicago: University of Chicago Press, 2000.
Johnston, Mark. *Saving God: Religion After Idolatry*. Princeton, NJ: Princeton University Press, 2009.
Kincaid, Harold, John Dupre, and Alison Wylie, eds. *Value-Free Science?: Ideals and Illusions*. Oxford: Oxford University Press, 2007.
Kisiel, Theodore. *The Genesis of Heidegger's "Being and Time."* Berkeley: University of California Press, 1995.
Kitcher, Philip. *Living with Darwin: Evolution, Design, and the Future of Faith*. Oxford: Oxford University Press, 2009.
Knauft, Bruce. *Genealogies for the Present in Cultural Anthropology*. New York: Routledge, 1996.
Köhnke, Klaus Christian. *Der Junge Simmel*. Frankfurt: Suhrkamp, 1996.
Krech, Volkhard. "From Historicism to Functionalism: The Rise of Scientific Approaches to Religions Around 1900 and their Socio-Cultural Context." *Numen* 47, no. 3 (2000): 244–65.
———. *Georg Simmels Religionstheorie*. Religion und Aufklärung, vol. 4. Tübingen: Mohr Siebeck, 1998.
Krell, David Farrell. *Daimon Life: Heidegger and Life-Philosophy*. Bloomington: Indiana University Press, 1992.
Kropotkin, Peter. *Mutual Aid: A Factor of Evolution*. Edited by Paul Avrich. New York: New York University Press, 1972.
Leiter, Brian. "Nietzsche's Metaethics: Against the Privilege Readings." *European Journal of Philosophy* 8, no. 3 (2000): 277–97.
———. "Nietzsche's Naturalism Reconsidered." In *The Oxford Handbook of Nietzsche*, edited by Ken Gemes and John Richardson. Oxford: Oxford University Press, 2013.
Lewens, Tim. *Darwin*. New York: Routledge, 2007.
———. "Human Nature: The Very Idea." *Philosophy and Technology* 25 (2012): 459–74.
Longino, Helen. *Science as Social Knowledge: Values and Objectivity in Scientific Inquiry*. Princeton, NJ: Princeton University Press, 1990.
Löwith, Karl. *Von Hegel zu Nietzsche: Der Revolutionäre Bruch im Denken des neunzehnten Jahrhunderts*. Stuttgart: Kohlhammer, 1950. Translated as *From Hegel to Nietzsche: The Revolution in Nineteenth-Century Thought*. New York: Columbia University Press, 1991.
May, Simon, ed. *Nietzsche's "On the Genealogy of Morality": A Critical Guide*. Cambridge: Cambridge University Press, 2011.
Millgram, Elijah. Critical notice of *Life and Action*, *Analysis Reviews* 69, no. 3 (2009): 557–64.
Moore, Gregory. *Nietzsche, Biology, and Metaphor*. Cambridge: Cambridge University Press, 2002.
Nagel, Thomas. *The Last Word*. Oxford: Oxford University Press, 1997.
———. *Mind and Cosmos: Why the Materialist Neo-Darwinian Conception of Nature Is Almost Certainly False*. Oxford: Oxford University Press, 2012.
———. *Secular Philosophy and the Religious Temperament: Essays 2002-2008*. Oxford: Oxford University Press, 2010.
———. *The View from Nowhere*. Oxford: Oxford University Press, 1989.
Nowak, Martin. "Five Rules for the Evolution of Cooperation." *Science* 314 (2006): 1560–63.
———, Corina Tarnita, and E. O. Wilson. "The Evolution of Eusociality." *Nature* 466 (2010): 1057–62.

Norenzayan, Ara. *Big Gods: How Religion Transformed Cooperation and Conflict.* Princeton, NJ: Princeton University Press, 2013.

———. Joseph Henrich, and Edward Slingerland. "Religious Prosociality: A Synthesis." In *Cultural Evolution: Society, Technology, Language, and Religion*, edited by Peter J. Richerson and Morten H. Christiansen. Cambridge, MA: MIT Press, 2013.

Ogden, Schubert. *On Theology.* Dallas, TX: Southern Methodist University Press, 1986.

Okasha, Samir. *Evolution and the Levels of Selection.* Oxford: Oxford University Press, 2008.

Peter, Niklaus. "Ernst Troeltsch auf der Suche nach Franz Overbeck: Das Problem des Historismus in der Perspektive zweier Theologen." *Troeltsch-Studien* 11 (2000): 94–122.

———. *Im Schatten der Modernität: Franz Overbecks Weg zur "Christlichkeit unserer heutigen Theologie."* Stuttgart: J. B. Metzler, 1992.

———. "Unerledigte Anfragen und befragte Erledigungen: Eine Erste Rezeption und Diskussion dreier Beiträge." In *Franz Overbecks unerledigte Anfragen an das Christentum*, edited by Rudolf Brändle and Ekkehard W. Stegemann, 196–210. Munich: Chr. Kaiser Verlag, 1988.

Pinkard, Terry. *Hegel's Naturalism: Mind, Nature, and the Final Ends of Life.* Oxford: Oxford University Press, 2012.

Pippin, Robert. *Hegel's Idealism: the Satisfactions of Self-Consciousness.* Cambridge: Cambridge University Press, 1989.

———. *Idealism as Modernism: Hegelian Variations.* Cambridge: Cambridge University Press, 1997.

———. "Natural and Normative." *Daedelus* 138, no. 3 (2009): 35–43.

Poellner, Peter. "Affect, Value, Objectivity." In *Nietzsche and Morality*, edited by Brian Leiter and Niel Sinhababu, 227–61. Oxford: Oxford University Press, 2007.

———. *Nietzsche and Metaphysics.* Oxford: Oxford University Press, 1995.

———. "Phenomenology and Science in Nietzsche." In *A Companion to Nietzsche*, edited by Keith Ansell Pearson, 297–313. Oxford: Blackwell, 2006.

Preus, J. Samuel. *Explaining Religion: Criticism and Theory from Bodin to Freud.* New York: Oxford University Press, 1996.

Putnam, Hilary. *Ethics without Ontology.* Cambridge, MA: Harvard University Press, 2004.

Richards, Robert. *The Romantic Conception of Life: Science and Philosophy in the Age of Goethe.* Chicago: University of Chicago Press, 2002.

Richardson, John. *Nietzsche's New Darwinism.* Oxford: Oxford University Press, 2004.

———. "Nietzsche on Life's Ends." In *The Oxford Handbook of Nietzsche*, edited by Ken Gemes and John Richardson. Oxford: Oxford University Press, 2013.

Robertson, Simon. "Normativity for Nietzschean Free-Spirits." *Inquiry* 54, no. 6 (2011): 591–613.

Rudwick, Martin. *Earth's Deep History: How It Was Discovered and Why It Matters.* Chicago: University of Chicago Press, 2014.

Santayana, George. *Skepticism and Animal Faith.* New York: Dover Press, 1955.

Schacht, Richard, ed. *Nietzsche, Genealogy, Morality.* Berkeley: University of California Press, 1994.

Schilbrack, Kevin. *Philosophy and the Study of Religions: A Manifesto.* Cambridge, MA: Wiley-Blackwell, 2014.

Schnädelbach, Herbert. *Philosophy in Germany, 1831–1933.* Translated by Eric Matthews. Cambridge: Cambridge University Press, 1984.

Scott-Phillips, Thomas, Thomas Dickins, and Stuart West. "Evolutionary Theory and the Ultimate–Proximate Distinction in the Human Behavioral Sciences." *Perspectives on Psychological Science* 6, no. 1 (2011): 38–47.

Slingerland, Edward. *What Science Offers the Humanities: Integrating Body and Culture.* Cambridge: Cambridge University Press, 2008.
Sober, Elliot and David Sloan Wilson. *Unto Others: The Evolution of Altruism.* Cambridge, MA: Harvard University Press, 1998.
Sommer, Andreas. *Der Geist der Historie und das Ende des Christentums: Zur 'Waffengenossenschaft' von Friedrich Nietzsche und Franz Overbeck.* Berlin: Akademie Verlag, 1997.
———. "Ein Philosophisch-Historicher Kommentar zu Nietzsches *Götzen-Dämmerung*: Probleme und Perspektiven." In *Perspektiven der Philosophie: Neues Jahrbuch*, vol. 35, edited by Georges Goedert and Martina Scherbel. Amsterdam: 2009.
———. "Nietzsche mit und gegen Darwin in den Schriften von 1888." In *Nietzsche, Darwin, und die Kritik der Politischen Theologie*, edited by Volker Gerhardt and Renate Reschke. Berlin: Akademie Verlag, 2010.
———. "Nietzsche und Darwin." In *Nietzsche als Philosoph der Moderne*, edited by Barbara Neymeyer and Andreas Sommer, 223–40. Heidelberg: Universitätsverlag, 2012.
———. "On the Genealogy of the Genealogical Method: Overbeck, Nietzsche, and the Search for Origins." *Bulletin of the Institute of Classical Studies.* Supplement Series 46, supp. 79 (2003): 87–103.
———. "Weltentsagung, Skepsis und Modernitätskritik. Arthur Schopenhauer und Franz Overbeck." *Philosophisches Jahrbuch* 107, no. 1 (2000): 192–206.
Stack, G. J. "Kant, Lange, and Nietzsche: Critique of Knowledge." In *Nietzsche and Modern German Thought*, edited by Keith Ansell Pearson. New York: Routledge, 1991.
Stearns, Stephen C., and Rolf F. Hoekstra. *Evolution: An Introduction.* 2nd ed. Oxford: Oxford University Press, 2005.
Sterelny, Kim, and Paul Griffiths. *Sex and Death: An Introduction to Philosophy of Biology.* Chicago: University of Chicago Press, 1999.
Street, Sharon. "A Darwinian Dilemma for Realist Theories of Value." *Philosophical Studies* 127, no. 1 (2006): 109–166.
Taylor, Charles. *A Secular Age.* Cambridge, MA: Harvard University Press, 2007.
Thompson, Michael. *Life and Action: Elementary Structures of Practice and Practical Thought.* Cambridge, MA: Harvard University Press, 2008.
Voigt, Friedemann. *Die Tragödie des Reiches Gottes? Ernst Troeltsch als Leser Georg Simmels.* Troeltsch-Studien, 10. Gütersloh: Gütersloher Verlag Haus, 1998.
Walsh, Denis. *Organisms, Agency, and Evolution.* Cambridge: Cambridge University Press, 2015.
Weber, Max. "Science as a Vocation." In *The Vocation Lectures*, edited by David Owen and Tracy Strong, translated by Rodney Livingstone. Indianapolis, IN: Hackett, 2004.
Williams, Bernard. "Nietzsche's Minimalist Moral Psychology." In *Nietzsche, Genealogy, Morality: Essays on Nietzsche's "On the Genealogy of Morals."* Edited by Richard Schacht, 237–47. Berkeley: University of California Press, 1994.
———. *Truth and Truthfulness: An Essay in Genealogy.* Princeton, NJ: Princeton University Press, 2002.
Woodford, Peter. "Neo-Darwinists and Neo-Aristotelians: How to Talk about Natural Purpose." *History and Philosophy of the Life Sciences* 38, no. 4 (2016): 1–23.
———. "The Very Possibility of a Science of Religion: Ernst Troeltsch and Neo-Kantianism," *Journal of Religion* 97, no. 1 (2017): 56–78.
Zachhuber, Johannes. *Theology as Science in Nineteenth-Century Germany: From F. C. Baur to Ernst Troeltsch.* Oxford: Oxford University Press, 2013.

Index

affects, 34–39, 86–87, 99, 118, 157n2
affirmation, 25, 33–51, 66–71, 76, 86–87, 111, 117, 121, 128, 133–45, 167n23, 169n1
agency, 4, 16–21, 23–24, 91–92, 104–8, 122–36, 150, 169n9. *See also* will
aims, 59–61, 63, 69, 101, 107–8, 114, 137–40
altruism, 6–10. *See also* evolution; group selection
Analytic philosophy, viii, 8, 109, 140
Anderson, R. Lanier, 86
animal faith, 169n1
animality, 140
Antichrist, The (Nietzsche), 37–38, 41, 49
"Anti-Darwin" (Nietzsche passage), 36–38
Antrittsvorlesung (Overbeck), 55–63
Apel, Karl-Otto, 140
a priori, 22, 80, 100, 105–8, 120–23. *See also* Kant, Immanuel
Aristotle, 22–23, 33, 141, 144, 160n58
asceticism, 44–51, 55, 66–67, 70–78, 128, 160n51
"Attempts at a Philosophy of Life" (Scheler), 13
autonomy (of values), 43–44, 63, 91–101, 105–8, 118–20, 130, 134, 155n29

Bachofen, Johann Jakob, 53
Baden school (of neo-Kantianism), 5, 16, 105–8, 167n36
Barth, Karl, 55
Baur, Ferdinand Christian, 53, 56–58, 68, 71
Being and Time (Heidegger), 107
Beiser, Frederick, 80, 163n3, 168n64
Benjamin, Walter, 55, 105
Bergson, Henri, 13, 79–80, 108, 112, 142–43, 155n24, 164–65n37, 166n14
Beyond Good and Evil (Nietzsche), 11, 37–38
Beyschlag, Willibald, 69–70, 76
Biologische Probleme zugleich als Versuch zur Entwicklung einer rationallen Ethik (Rolph), 36
biologism, 111–20, 124, 134–36, 140–43, 166n10
biology: Darwin and, 13–14; ethics and, 16, 39–41; historicism and, 52–63, 70–78; Lamarck and, 15–16; *Lebensphilosophie* and, 14–17; Nietzsche's relation to, 27, 29–33; teleology and, 21–24, 48–51. *See also* evolution; life; nature
Birth of Tragedy, The (Nietzsche), 25, 31–32, 34, 41
Burckhardt, Jacob, 33, 53, 59, 158n17

care, 126, 136, 147–48
Carnap, Rudolf, 140
Cassirer, Ernst, 106, 140–43
categorical imperative, 103
Christianity, 44–51; asceticism and, 55, 66–67, 70–78; genealogical explorations of, 55–63; modernity and, 52–55, 74–79; Nietzsche on, 41–47, 142; origins of, 52–53, 55–63; Schopenhauer on, 84–85; Simmel on, 84; vital interests of, 63–70, 76. *See also* religion
Clark, Maudemarie, 158n7
Clemens of Alexandria, 56
cognitive science, vii, 149
Cohen, Hermann, 9, 106
Continental philosophy, viii, 8, 109, 140
"Contributions to the Epistemology of Religion" (Simmel), 97
"Contribution to the Sociology of Religion, A" (Simmel), 97
creativity, 93–104, 117, 132–34, 155n24
Critique of the Power of Judgment (Kant), 22
culture, 1–11, 31–39, 48, 52–67, 71–104, 126, 134–35

Darwin, Charles: Aristotle and, 143–44; ethics and, 25–28; natural selection and, 2, 6–10,

Darwin, Charles (*continued*)
  36–37, 44, 111–12, 115, 144–45; neo-Darwinism and, 112, 153n6; Nietzsche and, viii, 22–23, 25–28, 35–39, 42–43, 48–51, 108, 145–46, 153n6; religion and, ix–x, 41–47; Rickert and, 108–18; seminal importance of, viii, 1–5, 11, 13, 134–35; in Simmel's work, 82–88, 94–97, 104
Dasein, 143
Davos disputation, 140–43
Dawkins, Richard, 9, 149
Der Alte und Der Neue Glaube (Strauss), 64, 66–67
Der Gegenstand der Erkenntnis (Rickert), 120
Der Junge Simmel (Köhnke), 81
desire, 30, 33–42, 84–85, 90–91, 104, 112, 132–34
dialectic, 40–41, 44, 46, 56–57, 68
Die Philosophie des Geldes (Simmel), 81–82, 87–94, 102
Die Philosophie des Lebens (Rickert), 108
Die Probleme der Geschichtsphilosophie (Simmel), 80–81
Dilthey, Wilhelm, 13, 105, 108, 141, 166n14, 167n34
disenchantment, 31, 76
Dole, Andrew, 155n29
Driesch, Hans, 14, 141
drives, 23–24, 30, 33–43, 48–51, 84, 86–87, 118, 133–34
Durkheim, Émile, 87

Ecce Homo (Nietzsche), 33–35, 38
economics, 87–92, 97–98, 102–4, 133–34, 165n52. See also *Die Philosophie des Geldes* (Simmel)
Einleitung in die Moralwissenschaft (Simmel), 81, 83, 103
élan vital, 112, 142
embryology, 141
empiricism, 8–9
"Empiricism and Platonism in the Philosophy of Religion" (Troeltsch), 155n24
Enlightenment, 10–15, 20, 54, 108
epigenetics, 156n30
epistemology, 31, 50, 82, 86, 97–98, 107, 111–26, 129–31, 144
eros, 30, 38, 50, 74–75, 84–85, 104, 112, 145
eternal recurrence of the same, 82–83
ethics: asceticism and, 44–51, 55, 66–78, 128, 160n51; economics and, 87–92, 97–104, 165n52; evolutionary pictures of, viii, 6–10, 43–47, 88–92, 154n11, 169n7; fact-value dichotomy and, 22, 55–56, 61–62, 79–91, 112–31; intrinsic value and, 16–21, 29–33, 45–47, 147; life's value and, 16, 28–35, 39–41, 132–34, 160n58; meta-ethics and, 8, 16, 149, 170n18; nature and, 3–5, 9, 76–78, 115–18, 145–46; Nietzsche and, viii, 25–28, 88; nontheoretical values and, 124–27; normativity and, 16–21, 30–31, 49–50; psychology and, 33–38, 42; rationalism and, 90–91; religion and, viii, 16–21, 41–47, 70–78, 98–104, 127–31, 146–49; science and, 16–21, 30, 47–51, 100–104, 118–20; teleology and, 21–24; transcendental theories of, 116–23, 130–31; validity problems and, 16–21, 73–97, 106–8, 110–18, 137–43, 147–48; value for its own sake and, 8, 39–41; vital interests and, 63–70
Eucken, Rudolf, 14, 166n16
evolution: creativity and, 93–104, 117, 132–34, 155n24; game theory and, 6–7; historical consciousness and, 4–5, 10, 30, 41–47, 52–63, 70–78, 80–81, 105, 147, 158n11; mortality and, 159n46; natural selection and, 2, 36–37, 44, 111, 144–45; of pro-social behaviors, 6–10; religion and, viii, 7–8, 69–70, 156n39; Simmel on, 94–97, 101, 104, 112; teleology and, 2, 37–39, 85–87, 139–40, 144–45; values and, viii, 6–10, 18–21, 43–47, 88–92, 114–15, 129–31, 154n11, 169n7. See also Darwin, Charles; life; natural selection
exchange, 92–93. See also *Die Philosophie des Geldes* (Simmel)
existentialism, 108, 132
eye, 22

facts (and values), 22, 55–56, 61–62, 79–91, 112–13, 115–18, 120–31
feelings of value, 81, 87, 97
Fichte, Johann Gottlieb, 15, 131
Foot, Philippa, 145–46, 170n12
forms of life, 13, 20, 41, 58, 89–92, 96–97, 135
Foucault, Michel, 26
freedom, 18, 45, 53–54, 62–63, 75–78, 102, 111. See also agency; autonomy (of values)
"Friedrich Nietzsche" (Simmel), 163n9

game theory (evolutionary), viii, 6–8, 156n39, 159n46
Gay Science, The (Nietzsche), 39
Geist, 11, 17, 24, 43, 49, 55–63, 68, 80, 83, 94–101, 107
genealogy, 21, 26–27, 55–63, 66–67, 76–77, 90–92, 99–100
Glauben, 17, 64, 72, 97, 104, 130–31
Goethe, Johann Wolfgang von, 2, 15, 108, 143
Gossman, Lionel, 53
Gould, Stephen Jay, 154n14, 162n45
Grene, Marjorie, 143–44
group selection, 159n46

Habermas, Jürgen, 140
Hamilton, William, 6
Hamilton's Rule, 154n9
Harnack, Adolf von, 65
Harris, Sam, 9, 149
Hartmann, Eduard von, 36

Hegel, G. W. F., 10–11, 15–17, 53, 56–58, 68, 71, 80, 120, 143, 154n16
Heidegger, Martin, 28–29, 55, 105, 107–8, 140–43, 168n63
Henry, Martin, 161n8
Herder, Johann Gottfried, 2, 143
Herodotus, 60
historicism, 10, 30, 41–47, 52–63, 70–78, 80–81, 105, 147, 158n11
Historicism and Its Problems (Troeltsch), 164n11
historiography, vii–ix, 6–10
History of Greek Culture (Burckhardt), 158n17
Hitchens, Christopher, 9, 149
Human, All Too Human (Nietzsche), 36
Humboldt, Alexander von, 2–3, 143
Hume, David, 8, 15, 42, 155n28
Hursthouse, Rosalind, 145
Hussain, Nadeem, 49–50, 158n35
Husserl, Edmund, 9, 105, 141–42, 167n34

idealism, 10, 16–17, 24, 28, 68–69, 84, 120–27, 143
ideal of life, 63–70, 161n7
ideal worlds, 94–97, 99–100, 102–4
immanence, 99–100
individual law, 82, 103–4, 111
infinite totality, 127
instrumentality, 16–21, 88, 98, 129, 145
intellectualism, 125–27
intentionality, 141
intuitionism, 108, 111–12, 166n14
Irenaus, 56
irrationalism, 10, 108, 130–31

Jacobi, Friedrich Heinrich, 14–15, 108, 155n28
James, William, 13, 106, 155n24, 166n13
Janaway, Christopher, 157n2
Jaspers, Karl, 105, 143–44
Joas, Hans, 16, 155n24
Judaism, 66–67
judgment, 91–92, 94–97, 112–13, 115, 120–23, 148, 157n2, 168n63

Kant, Immanuel, 8–9, 15–17, 22–23, 54, 79–84, 103–8, 119, 126, 129, 143, 148
Kant and Goethe (Simmel), 82
Kirchenlexicon (Overbeck), 54, 70–78
Klages, Ludwig, 13–14
Knauft, Bruce, 157n5
Köhnke, Klaus Christian, 81
Köselitz, Heinrich, 28
Krauss, Lawrence, 9, 149
Kropotkin, Peter, 6

Lagarde, Paul de, 64–65
Lamarck, Jean-Baptiste, 15–16, 156n30
Lange, Friedrich Albert, 36, 143, 156n34

Lask, Emil, 105, 168n63
Lazarus, Moritz, 80
Lebensanschauung (Simmel), 82, 94–97
Lebensideal, 63–73, 161n7
Lebensphilosophen. See life; philosophy; *and specific philosophers*
Lebensphilosophie (Rickert), 36
Leibniz, Gottfried Wilhelm, 166n16
Leiter, Brian, 158n7
Lessing, Theodor, 14
liberalism, 53–55, 62, 78
life: asceticism and, 44–47, 77–78; creativity and, 93–104, 117, 132–34, 155n24; desire for, 30–43, 45–46, 84–85, 112, 132–33; drives and, 23–24, 30, 33–43, 48–51, 84, 86–87, 118, 133–34; meanings of, 35–39, 49–51, 124–27; metaphysics of, 94–97, 99–100; more-than-life and, 94–97, 101, 104, 112, 133–34; nihilism and, 15, 29–33, 108, 117, 132, 155n28; normativity and, 112–18; Overbeck's notion of, 63–70, 161n7; political economy and, 87–93; religion and, 59–70, 84, 97–104, 161n7; Rickert's understanding of, 108–10; as self-desiring, 85–87; teleology and, 30–33, 85–87, 137–40, 158n31; valuation and, 39–41, 43, 70–78, 85–93, 109–18, 128, 132–34, 136–37, 158n35, 167n23, 169n1
logicism, 10
Lotze, Hermann, 16, 85–86, 105–6, 143, 156nn30–31
love, 74–75
Löwith, Karl, 55

Malthus, Thomas, 6, 111
Marburg school (of neo-Kantianism), 106, 167n36
Marx, Karl, 87, 89
materialism, 10, 134–37, 140–41
medicalization (of descriptive terms), 115–18, 133–35, 145, 158n9
meta-ethics, 8–9, 16, 111, 149, 170n18
Methodenstreit, 11, 107–10, 112–13
money, 81, 87–92, 97–104, 133–34, 165n52
Moore, G. E., 8–9
Moore, Gregory, 37, 158n9
moral imperative, 82
morality. See ethics; values
moral realism, 122–23, 130–31
more-than-life, 94–97, 101, 104, 112, 133–34
myth, 60, 65–66

Nagel, Thomas, 19–20, 28, 48, 153n6
Nägeli, Carl von, 29, 37
Natorp, Paul, 9, 106
Natural Goodness (Foot), 170n12
natural histories, 41–47, 126–27
Natural History of Religion (Hume), 42
naturalism, 108, 166n20, 170n16
naturalistic fallacy, 8–9, 111, 123

natural selection, 2, 6–10, 36–37, 44, 111–12, 115, 144–45
nature: creativity and, 93–104, 117, 132–34, 155n24; culture and, 1–11, 31–39, 48, 52–67, 71–104, 126, 130–31, 134–35; drives originating in, 4–5; ethics and, 16–21, 27–28, 39–47, 50–51, 77–78, 145–46; evolutionary theory and, viii, 1–5; history and, 4–5, 57–58, 70–78; intrinsic value and, 16–21; Kant on, 81; meaning of, x; as metaphysical foundation, 103–4; normativity and, 30–31, 49–50, 60–61, 76–87, 94–100, 106–23, 133–43, 166n21; religion and, 41–51, 56–63, 71–78, 80–82, 112, 132, 146–49; representations of, 137–40; Rickert's understanding of, 105–8; science and, 10–13; social psychology and, 5, 80–82; teleological interpretations of, 1–5, 10, 16–21, 143–44; theology and, 1–2; values originating in, 7–10, 39–41, 79–93, 112–13, 129–31; the will and, 16–21; will to power and, 28–29, 33–35, 37–38, 159n31. *See also* biologism; evolution; life; philosophy
neo-Kantianism, 5, 9, 16–17, 21, 36, 50, 79–90, 104–8, 129–35, 140–43, 156n34. *See also specific philosophers*
neuroscience, vii, 149
New Atheists, 9, 149
Nietzsche, Elisabeth Förster, 28
Nietzsche, Friedrich: academic career of, 12, 32; breakdown of, 4; Darwin's influence on, viii, 22–23, 25–28, 35–39, 42–43, 48–51, 108, 145–46, 153n6; eternal return and, 82–83; ethics and, 25–28, 88; Foot on, 146; Heidegger on, 28–29; Kant and, 8–9; life concept of, 5, 13, 28–33, 52–53, 59–60, 66–67, 76, 132–33, 164n41, 169n1; nature and, 4, 142–43, 158n7; nihilism and, 29–33, 155n28; Overbeck's relationship with, 52, 55, 59, 64–65, 71–72; philosophical influence of, viii; rationalism and, 15; religion and, 41–47, 97–98, 136, 142; Rickert and, 106, 108–10, 115–16, 127–28, 134–36, 167n23; Schopenhauer and, 15–16; Simmel on, 82–93, 97, 102–4, 133–34, 164n11; tragedy and, 31–35, 38, 59, 96–97, 102–4; will to power and, 28–29, 37–38, 159n31. *See also specific works*
"Nietzsche und Kant" (Simmel), 163n9
nihilism, 15, 29–33, 108, 117, 132, 155n28
normativity, 30–31, 49–50, 60–61, 76–87, 94–100, 106–23, 133–43, 166n21
Nowak, Martin, 154n9

objective spirit, 80–81
Ogden, Schubert, 20, 38–39, 48
"On the Emergence and Right to a Purely Historical Consideration of the New Testament Scriptures in Theology" (Overbeck), 55–63

On the Genealogy of Morality (Nietzsche), 23, 41–42, 44–47, 67
"On the Personality of God" (Simmel), 97
"On the Use and Disadvantage of History for Life" (Nietzsche), 30
Origen, 56
Ortega y Gasset, José, 13–14
Overbeck, Franz, 12–13, 15–16, 133; academic career of, 52–55, 70–71, 76, 161n8; Nietzsche's relationship with, 4, 30, 52, 55, 59, 64–65, 71–72, 160n51; Schopenhauer and, 15–16; vital interests and, 53–55, 63–70, 133, 147. *See also specific works*

Paley, William, 1
perfection principle, 37
pessimism, 15, 33–34, 74, 86–87
Pflege, 126, 136, 147–48
phenomenology, 92, 106–7, 120, 134, 140–43, 167n34. *See also specific philosophers*
Phenomenology of Spirit (Hegel), 120
philosophy: Analytic-Continental divide in, viii, 8, 109, 140; historiography and, vii–ix, 6–10; knowledge for its own sake and, 20–21; methodological rigor and, 106–10, 112–13; nature and, 20–21; positivism and, 8–9; religion and, viii, 2–3, 146–49; science's relation to, vii, viii–ix, 10–13, 50–51, 85–86, 105–18, 120–23, 149. *See also* life; nature; religion; science; *and specific philosophers*
"Philosophy of Fashion, The" (Simmel), 82
Philosophy of Life, The (Rickert), 5, 108–10
Philosophy of Money, The (Simmel). *See Die Philosophie des Geldes* (Simmel)
Plessner, Helmuth, 141
Poellner, Peter, 158n7
Poetics (Aristotle), 33
positivism, 8–10, 80–81, 105, 146–49
pragmatism, 10, 13, 97–98, 106, 115, 148, 166n13, 166n20, 170n16
presentism, 109–10
Principia Ethica (Moore), 8
"Problem in the Philosophy of Religion, A" (Simmel), 97
"Problem of the Religious Situation, The" (Simmel), 97
Protestantism and science thesis, 61–63, 77–78
psychologism, 10, 83, 85–86, 124, 167n35
psychology, 5–10, 26, 33–35, 37–38, 42, 45, 47–51, 79–82, 90–92, 159n42

quietism, 54, 117

rationalism, 9, 15, 24–47, 56–57, 68, 74, 83, 90–91, 104–8, 120–40, 154n16, 167n23
reductionism, 4, 89

Reformation, 56, 61–62
regulative principles, 107–8, 134–35
relativism, 88–90
religion: aims and interests of, 55–63, 126–31; asceticism and, 44–51, 55, 66–67, 70–78, 128, 160n51; Darwin's theses and, ix–x; ethics and, 16–21, 41–47, 70–78, 98–104, 127–32, 146–49; evolutionary pictures of, viii, 7–8, 156n39; historical consciousness and, 4–5, 52–63; life's value and, 16, 63–70, 77–78, 84, 98–104, 112, 161n7; modernity and, 52–55, 64–70; natural histories and, 41–47, 71–78, 80–82, 132, 146–49; Nietzsche on, 41–47, 136; philosophy of, viii, 2–3, 146–49, 170n15; science's relation to, 9–13, 16–17, 27–28, 42, 47–52, 59–82, 100–104, 146–50, 154n14, 162n45; teleology and, 2, 18–21; theology and, 1–2, 161n8; values expressed in, 9–10, 16–17, 168n56; vital interests and, 64–70
"Religion" (Simmel), 82, 97
representation, 120–23, 137–40, 147
ressentiment, 43, 142
revaluation of values, 25–28, 43–47, 49, 67–68, 102, 145
Richards, Robert, 153n5
Richardson, John, 157n6, 158n7
Rickert, Heinrich, 5, 9–21, 24, 28, 36, 50, 79–80, 85–86; biologism and, 110–18, 134, 166n10; Nietzsche's appraisal by, 108–10, 115–16, 127–28, 167n23; religion and, 147–48, 168n56; science as task and, 122–27, 139–40; Simmel on, 108, 122, 125, 142; theory of value of, 120–23, 147, 154n7, 167n47. *See also specific works*
Rolph, William, 29, 36–37
Romanticism, 2–3
Roux, Wilhelm, 29, 37

Santayana, George, 169n1
Schacht, Richard, 158n7
Scheler, Max, 13, 28, 108, 141–42
Schelling, Friedrich Wilhelm, 2, 143, 155n29
Schilbrack, Kevin, 20
Schlegel, Friedrich, 14
Schleiermacher, Friedrich, 2, 143, 155n29
Schnädelbach, Herbert, 16
Schopenhauer, Arthur, 14–16, 34, 72, 82–85, 108, 143, 163n3
Schopenhauer and Nietzsche (Simmel), 82–83
Schopenhauer und Nietzsche (Simmel), 163n9
science: asceticism and, 47–51; ethics and, 7–8, 16–21, 41–51; historicism and, 55–63; knowledge for its own sake and, 16–21, 30, 47–51, 54–55, 57–58, 65–66, 68, 70, 95, 97–101, 109–27, 133–35, 137, 139–40, 145, 147, 150, 170n15; life's meaning and, 124–27; mythology and, 60–61; *Natur-Geist* split in, 11, 17, 24, 43, 49, 55–63, 94–101, 107; normativity and, 97–98; philosophy's

relation to, vii, viii–ix, 10–13, 50–51, 85–86, 105–18, 120–23, 144, 149; Protestantism and, 61–63; rationalism and, 33–35; reductionism and, 4; religion's relation with, 9–13, 16–17, 27–28, 42, 47–52, 54–55, 59–82, 97, 100–104, 130–31, 146–50, 154n14, 162n45; as task, 121–23; teleology and, 21–24, 30–33; theology and, 59–70, 74; tragedy and, 31–32; values of, 9–10, 110–23, 132
Simmel, Georg, 12, 30, 50, 164n39; fact-value distinction and, 79–82; *Lebensphilosophie* and, 13; monetary value and, 87–93, 141; Nietzsche's uptake by, 5, 82–93, 102–4, 133–34, 163n5, 164n11; religion and, 97–104, 168n56; Rickert on, 108, 122, 125, 142; Schopenhauer and, 15–16; *Sein/Sollen* dichotomy and, 92–97. *See also specific works*
skepticism, 117, 122
Slingerland, Edward, 170n18
social psychology, 5–10, 26
Sockness, Brent, 166n16
Socrates, 32, 34, 39–41, 43–44, 46, 68
Sommer, Andreas, 36–37, 162n22
Southwest school (of neo-Kantianism), 5, 16, 105–8, 167n36
specialist universalism (fallacy), 114
Spencer, Herbert, 13–14, 108, 111
Spengler, Oswald, 13–14
Spinoza, Baruch, 2, 14–15, 155n28, 164n27
Spir, African, 156n34
Steinthal, Heymann, 80
Strauss, David Friedrich, 64–65, 67, 77
symptomology, 39, 50–51
System of Philosophy (Rickert), 122

Taine, Hippolyte, 87
teleology: agency and, 4, 21, 35–39, 124–27; Aristotelian, 143–44; ethics and, 16–21; historical consciousness and, 55–63; nature and, 1–2, 10, 16–21, 48–51, 85–87, 137–44; Nietzsche and, 29–33, 132; as regulative principle, 22–23; science and, 121–27, 133–35
temporality, 60–61, 74–76
Tertullian, 56
theology, 161n8; evolution and, 69–70; Heidegger and, 55; historical consciousness and, 4–5, 52–63, 76; natural, 1–2; Overbeck and, 4; science and, 59–70, 74, 77–78. *See also* religion
Thompson, Michael, 145–46
Thucydides, 60–61
tragedy, 31–35, 38, 59, 96–97, 102–4
transvaluation (of values). *See* revaluation of values
Trendelenburg, Friedrich, 106, 143
Troeltsch, Ernst, 55, 70–71, 105, 154n15, 155n24, 164n11, 168n63
truth (as value), 110–20, 124–27

Truth and Truthfulness (Williams), 60
Twilight of the Idols (Nietzsche), 33–34, 36, 39–40

Ueber die Christlichkeit unserer heutigen Theologie (Overbeck), 54, 63–70, 161n12
Uexküll, Jacob von, 141
Unamuno, Miguel de, 13
University of Basel, 4, 32–33, 53, 55–56, 59
Untimely Meditations (Nietzsche), 30–32, 37, 47–48, 59, 64–65

validity, 16–21, 26–27, 73–97, 106–8, 110–27, 129–31, 133–35, 137–43, 145, 147–48
values: autonomous region of, 43–44, 63, 91–101, 105–8, 118–20, 130, 134, 155n29; evolution and, viii, 6–10, 43–47, 88–92, 154n11, 169n7; fact-value divide and, 22, 55–56, 61–62, 79–82, 85–87, 89–91, 112–13, 115–18, 120–31; feelings of, 81, 87, 97; intrinsic, 8, 16–21, 29–33, 39–41, 45–47, 147; life as ground of, 3–5, 9, 16, 28–35, 39–41, 76–78, 115–18, 132–34, 145–46, 160n58; monetary, 87–92, 97–104, 133–34, 165n52; normativity and, 30–31, 49–50, 60–61, 76–87, 94–100, 106–23, 133–43, 166n21; revaluation of, 25–28, 43–47, 49, 67–68, 102, 145; of science, 16–21, 30, 47–51, 100–104, 118–27, 133–35; validity and, 16–21, 73–97, 106–18, 137–43, 147–48
vital interests, 52–53, 64–70, 74–75, 77–78, 112, 133, 136
vitalism, 10, 66, 108, 112–14, 119, 140–42, 156n31
Völkerpsychologie, 5, 80–83
"Vom Begriff der Philosophie" (Rickert), 124

Weber, Max, 70, 87, 105, 146–47, 170n14
Wertphilosophie, 80–81, 83, 107–9
Whewell, William, vii
Whitehead, Alfred North, 143–44
Wilde, Oscar, 13
will, 15–21, 33–35, 37–38, 46–47, 84–85, 158n31
Williams, Bernard, 60–61, 65, 68, 157n2
will to power, 28–29, 37–38, 159n31
Will to Power, The (Nietzsche), 28–29
Windelband, Wilhelm, 9, 11, 16, 85–86, 105–8, 156n30
Wissen. See *Geist*; *Glauben*; science
worldviews, 10–13, 38, 60–70, 82–83, 97–108, 124–27, 137–40, 168n50
Wundt, Wilhelm, 167n34

"Zum Verständnis Nietzsches" (Simmel), 163n9

Printed and bound by CPI Group (UK) Ltd, Croydon, CR0 4YY
09/06/2025

14685712-0001